Debating the D

Since President Nixon coined the phrase, the "War on Drugs" has presented an important change in how people view and discuss criminal justice practices and drug laws. The term evokes images of militarization, punishment, and violence, as well as combat and the potential for victory. It is no surprise then that questions such as whether the "War on Drugs" has "failed" or "can be won" have animated mass media and public debate for the past 40 years.

Through analysis of 30 years of newspaper content, *Debating the Drug War* examines the social and cultural contours of this heated debate and explores how proponents and critics of the controversial social issues of drug policy and incarceration frame their arguments in mass media. Additionally, it looks at the contemporary public debate on the "War on Drugs" through an analysis of readers' comments drawn from the comments sections of online news articles.

Through a discussion of the findings and their implications, the book illuminates the ways in which ideas about race, politics, society, and crime, and forms of evidence and statistics such as rates of arrest and incarceration or the financial costs of drug policies and incarceration are advanced, interpreted, and contested. Further, the book will bring to light how people form a sense of their racial selves in debates over policy issues tied to racial inequality such as the "War on Drugs" through narratives that connect racial categories to concepts such as innocence, criminality, free will, and fairness. *Debating the Drug War* offers readers a variety of concepts and theoretical perspectives that they can use to make sense of these vital issues in contemporary society.

Michael L. Rosino is Assistant Professor of Sociology at Molloy College. His research and teaching focus on racial politics, media, social movements, crime, law and deviance, and human rights. His work emphasizes social change, policy, and community and civic engagement. He has published widely on the connections between racial oppression, struggles for racial equality, political conflicts, debates over public policy, and everyday social life in various scholarly and public outlets. His current research examines how activists within progressive grassroots political organizations engage with racial and political inequality through their identities, habits, and political strategies. The project illuminates the possibilities and barriers for building a racially just and inclusive grassroots democracy and advances new understandings of racial politics grounded in everyday social life.

Framing 21st Century Social Issues
Series Editor: France Winddance Twine, University of California, Santa Barbara

The goal of this new, unique series is to offer readable, teachable "thinking frames" on today's social problems and social issues by leading scholars. These are available for view on http://routledge.customgateway.com/routledge-social-issues.html.

For instructors teaching a wide range of courses in the social sciences, the Routledge *Social Issues Collection* now offers the best of both worlds: originally written short texts that provide "overviews" to important social issues *as well as* teachable excerpts from larger works previously published by Routledge and other presses.

As an instructor, click to the website to view the library and decide how to build your custom anthology and which thinking frames to assign. Students can choose to receive the assigned materials in print and/or electronic formats at an affordable price.

Available:

Debating the Drug War
Race, Politics, and the Media
Michael L. Rosino

Changing Times for Black Professionals
Adia Harvey Wingfield

The Problem of Emotions in Societies
Jonathan H. Turner

Rapid Climate Change
Causes, Consequences, and Solutions
Scott G. McNall

Waste and Consumption
Capitalism, the Environment, and the Life of Things
Simonetta Falasca-Zamponi

The Future of Higher Education
Dan Clawson and Max Page

Contentious Identities
Ethnic, Religious, and Nationalist Conflicts in Today's World
Daniel Chirot

Empire Versus Democracy
The Triumph of Corporate and Military Power
Carl Boggs

The Stupidity Epidemic
Worrying About Students, Schools, and America's Future
Joel Best

Sex, Drugs, and Death
Addressing Youth Problems in American Society
Tammy Anderson

Body Problems
Running and Living Long in a Fast-Food Society
Ben Agger

The U.S. Immigration Debate
The Myths and Realities of Immigration in the United States
Greg Prieto

Social Problems, 2nd Edition
A Human Rights Perspective
Eric Bonds

Series Advisory Board: Rene Almeling, *Yale University*, Joyce Bell, *University of Pittsburgh*, Elizabeth Bernstein, *Barnard College*, David Embrick, *University of Connecticut*, Tanya Golash-Boza, *University of California – Merced*, Melissa Harris, *New York University*, Matthew Hughey, *University of Connecticut*, Kerwin Kaye, *SUNY– Old Westbury*, Wendy Moore, *Texas A&M*, Alondra Nelson, *Columbia University*, Deirdre Royster, *New York University*, Zulema Valdez, *University of California – Merced*, Victor Rios, *University of California – Santa Barbara*.

Debating the Drug War

Race, Politics, and the Media

Michael L. Rosino

NEW YORK AND LONDON

First published 2021
by Routledge
52 Vanderbilt Avenue, New York, NY 10017

and by Routledge
2 Park Square, Milton Park, Abingdon, Oxon, OX14 4RN

Routledge is an imprint of the Taylor & Francis Group, an informa business

© 2021 Taylor & Francis

The right of Michael L. Rosino to be identified as author of this work has been asserted by them in accordance with sections 77 and 78 of the Copyright, Designs and Patents Act 1988.

All rights reserved. No part of this book may be reprinted or reproduced or utilised in any form or by any electronic, mechanical, or other means, now known or hereafter invented, including photocopying and recording, or in any information storage or retrieval system, without permission in writing from the publishers.

Trademark notice: Product or corporate names may be trademarks or registered trademarks, and are used only for identification and explanation without intent to infringe.

Library of Congress Cataloging-in-Publication Data
Names: Rosino, Michael L., author.
Title: Debating the drug war: race, politics, and the media / Michael L. Rosino.
Description: New York, NY: Routledge, 2021. |
Series: Framing 21st century social issues series
Identifiers: LCCN 2020042447 | ISBN 9781138239685 (hardback) | ISBN 9781138239692 (paperback) | ISBN 9781315295176 (ebook)
Subjects: LCSH: Drug control—United States. | Drug abuse—Government policy—United States. | Crime and race—United States. | United States—Race relations.
Classification: LCC HV5825 .R674 2021 | DDC 364.1/77—dc23
LC record available at https://lccn.loc.gov/2020042447

ISBN: 978-1-138-23968-5 (hbk)
ISBN: 978-1-138-23969-2 (pbk)
ISBN: 978-1-315-29517-6 (ebk)

Typeset in Sabon
by codeMantra

Contents

1	Introduction	1
2	The War on Drugs as a Contested Social Issue	21
3	How the Media "Frames" the Debate	42
4	Debate Dynamics: Racial Silence, Resonance, and Code Words	66
5	Identity Construction in the Heat of Debate	84
6	Conclusion	122
	References	145
	Glossary	169
	Index	181

Chapter 1
Introduction

On February 2, 2012, a plainclothes New York City police officer shot and killed Ramarley Graham, an unarmed 18-year-old black man. The shooting occurred in Graham's home, which he shared with his grandmother and six-year-old brother (Flegenheimer and Baker 2012). Ramarley was a student at the Young Scholars Academy of the Bronx with hopes of world travel and becoming a veterinarian (White 2017). Known as "Marley" to his family and friends, he liked to cook, play video games, and spend time with his older sister (Democracy Now! 2012). He was also one of three black men killed by the New York Police Department (NYPD) in that week alone (Goldstein 2012).

The United States is a nation indelibly shaped by an ongoing legacy of police brutality, racial profiling, and social injustices. In that sense, Graham's death is part of a national pattern of anti-black police violence. Yet, what is surprising in this case is the rationale employed by the NYPD to justify following and detaining him. After seeing him purchasing a small amount of cannabis on the street via surveillance cameras, officers followed him and entered his home without a warrant.

Many of the details remain murky. Yet, what is known is deeply troubling. Officer Richard Haste announced that Graham had a firearm, perhaps misrecognizing the young man adjusting the waistband of his pants (Mathias 2014). After breaking through the backdoor and entering the home, officers then broke down the bathroom door. Graham had gone into the bathroom to flush the cannabis down the toilet right before he was killed (Mathias 2014). The police never found a gun as Graham lay dead on the bathroom floor. After she'd experienced her grandson's traumatizing death, the police detained his grandmother for hours and questioned her about what she had witnessed, without giving her access to an attorney (Mathias 2014). Although the city settled a $3.9 million wrongful death suit with the family, the federal government or a grand jury failed to indict the officers (WNYC 2016).

The police killing of Ramarley Graham is one of many high-profile events that highlight the use of drug laws to defend policing practices that primarily target poor black and Latinx communities. The dominant methods of drug law enforcement in the US support surveillance, police brutality and violence, racial profiling and discrimination, and mass incarceration. These problems are linked to punitive drug control policies and militaristic policing practices – commonly known as the **War on Drugs** (Provine 2007; Alexander 2012). The War on Drugs formally began in the 1970s in the United States but, as we will see, these policies and practices build upon centuries-long legacy.

The US public increasingly supports reforms to decrease the harshness of drug policies, and many states have followed suit by legalizing or decriminalizing forms of drug possession and usage, particularly cannabis (also referred to as marijuana) (Ingram 2017). However, drug policy continues to elicit contestation. A host of conflicting narratives and perspectives about the purpose and impacts of drug law enforcement, and even why such reforms are important, sit just beneath the surface of this seeming consensus. Accordingly, the motivations and consequences of our drug prohibition policies and practices are frequent topics of discussion in print and digital media. These public discussions make up the **War on Drugs debate**.

As observed by journalist Dan Baum (1996:xi), "the War on Drugs is about a lot of things but only rarely is it really about drugs." The public debate over the War on Drugs provides a platform for people to talk about the nature of racial inequality and racism, political divides and the ideas of political opponents, the causes and demographics of crime, the role of police and prisons in society, ideals such as freedom and justice, and even how the government should use tax revenue and public resources. This book focuses on public discussions surrounding the War on Drugs that take place in mass media such as newspapers and the internet. However, given all these connected issues, approaching these debates from a sociological perspective can tell us about a lot more. What can we learn about race, politics, and the media from the debate over the War on Drugs? Let's start with some background.

The War on Drugs

On June 17, 1971, in a speech to Congress, President Richard Nixon famously argued for a "full-scale attack on the problem of drug abuse in America." He called for a national "war against heroin addiction." News outlets remarked on Nixon's militaristic rhetoric with headlines proclaiming variations of the phrase "War on Drugs." The following day, *The Chicago Tribune* announced: "Nixon Declares War on Narcotics Use in US" (Young 1971). Even outside of the US, the warlike imagery within this now infamous speech made headlines. London-based newspaper *The Guardian* ran the headline: "Nixon declares war on drug addicts" (Davenport 2011). Since then, the term "War on Drugs" has taken on a life of its own within mass media and everyday talk.

The phrase "War on Drugs" represents a crucial change in how Americans view and discuss criminal justice practices and drug laws. Despite the use of words like "attack," "war," and "strike" and a call for police to "further tighten the noose around the necks of drug peddlers," Nixon's speech mainly focused on treatment strategies to address heroin use among **Vietnam War** (1955–1975) veterans. Yet, in subsequent years, the phrase and its implications became translated into an actual set of warlike policies and practices (Alexander 2012).

The War on Drugs is both literal and metaphorical. It evokes images of militarization, punishment, and violence as well as strategy, combat, and the potential for victory. Like other wars, Americans have often questioned whether it has failed or can even be won at all. US Government spending on these policies and practices has been a combined total of over $1 trillion (Drug Policy Alliance 2017). So, like other military endeavors, the War on Drugs is often debated on the merits of this massive investment of public funds. These types of questions have animated public debate for the past 40 plus years and become a prominent topic in the media.

How did the War on Drugs become a war? National and even international governments such as the United Nations commonly draw distinctions between substances which provide legitimate medical benefits to their user and those which either do not have benefits or cause harm (Roberts and Chen 2013). These distinctions may seem objective or even natural. Yet, as Howard S. Becker (1963), a sociologist, pointed out, they are the products of people undertaking a **moral enterprise** to create rules and have them enforced. Governments that have deemed specific drugs harmful routinely also make them illegal to possess, sell, or consume under the guise of protecting individuals and communities. But the labeling of drugs and drug policies as either beneficial or harmful is a complicated assessment.

The existence of rules about what substances to consume and not consume is highly prevalent among human societies. However, the distinction between categories such as "drugs" and "medicine" varies by time and place. Whether the government and other institutions classify a chemical substance as a harmful drug or as beneficial medicine and whether the users of that substance are considered a threat to society or themselves is the result of political and social processes rather than objective standards (Becker 1963; Goode 2008). For instance, in the United States and many other countries, heroin is widely accepted as a dangerous and addictive substance with no medicinal benefits. However, Bayer Pharmaceuticals, the same company that introduced Aspirin, originally manufactured this substance as an analgesic or medicine used to alleviate pain (Goode 2008). Although the federal government now classifies heroin as a harmful drug with no medical benefit in the United States, its use continues in medical contexts in countries such as the United Kingdom (Goode 2008).

While the War on Drugs is a product of the 1970s and 1980s, the history of drug prohibition in the United States dates back over 140 years. In 1875, the city of San Francisco passed the nation's earliest drug prohibition. This ordinance made it a crime to run or visit opium dens (Fisher 2014; Garner 2014). Racism toward Chinese immigrants helped fuel growing fears about opium addiction in the United States. A November 19, 1875, article in the *Los Angeles Herald* stated, "San Francisco is inaugurating measures for the suppression of the opium dens established and frequented by the Chinese of that city. Those Mongolian death pens are no doubt great promoters of vice and disease." Despite the nonracial language in the actual ordinance, police enforcement focused on preventing whites from entering opium dens operated by Chinese immigrants (Fisher 2014). As a December 7, 1875, article in the

Los Angeles Herald reported about the San Francisco opium den prohibition:

> The first raid under the new ordinance against white persons who frequent opium dens was made at 12 o'clock this morning by a force of police under Captain Douglass and Detective Rogers. [...] The police intend to continue their raids until the growing evil is suppressed.

Within the decade, opium bans spread across the nation, and in states such as Iowa, laws overtly banned whites from opium dens.

These earliest laws reflected the growing perception that opium dens represented dangerous temptations for otherwise morally upstanding middle-class whites (Fisher 2014). A news item describing opium dens in Chico, California, appearing in the *Weekly Butte Record* published March 3, 1877, exemplifies these fears, claiming:

> [...] these narcotic dens are made places of resort by whites of both sexes, whose infatuation with opium is so great as to debase all their decent sensibilities and render them wholly impervious to shame, insensible to the shafts of ridicule and disregardful of the importunate pleadings of shocked friends. In the dens alluded to, Chinese and whites, male and female, gather together and worship the god of opium on the plane of perfect equality – their sole object being to get drunk and stupify themselves by inhaling the fumes of the narcotic poison.

Word of mouth and mass media reports raised concerns that white men and women might engage in lascivious and sinful behavior such as promiscuous sex while under the influence of Chinese opium (Ahmad 2000; Fisher 2014). Subsequent police efforts to crack down on opium dens in New York City took on even more explicitly racial motivations. A series of further reports stoked whites' animosity toward Chinese immigrants and anxieties over the potential for the so-called "race-mixing" between white women and Chinese men in these establishments (Ahmad 2000; Garner 2014). The fears that helped motivate these bans were part of a larger racial narrative depicting Eastern nations as a threat, later termed "yellow peril." These sentiments eventually led to the Chinese Exclusion Act of 1882, the first law to exclude an entire ethnic group from entry into the United States (Yang 2004).

Four decades later, prejudice and fear toward ethnic and religious groups helped produce another set of drug laws prohibiting alcohol. Anti-Catholic sentiment by groups such as the **Ku Klux Klan**, a white supremacist terrorist organization founded by Confederate soldiers after the Civil War in 1866 in Tennessee that experienced a resurgence in the 1920s, and rising antisemitism against Jewish immigrants involved in the alcohol trade bolstered demand for banning alcohol production and consumption (Cohen 2006; Marni 2012). In the early 20th century, Catholicism and Judaism were seen by many American Protestants as connected to distinct ethnic groups outside the bounds of the racial category of "white." These movements against alcohol became part of the social, economic, political, and legal conflicts, whereby various ethnic, cultural, and religious groups became categorized as "white" or "not quite white" (Roediger 1991; Jacobson 1999; Haney López 2006). Popular media and influential commentators depicted European migrants with these religious identities as threatening the dominant social and economic standing of white Protestants in the United States.

In the period leading up to prohibition, groups pushing narratives about the dangers of alcohol also associated it with Mexican immigrants. As noted by Marie Sarita Gaytán (2013:444), a sociologist, "Mexican men were frequently described in newspapers as drunk, ignorant, or savage," and police officers routinely gave them much harsher punishments for public intoxication than European Americans. Drinking establishments called saloons became increasingly regarded by those opposed to alcohol use as places where blacks, foreign immigrants, and working poor whites engaged in immoral behavior (Provine 2007; Muhammad 2011). In the South, temperance movements that sought to reduce alcohol consumption also associated alcohol with criminality among blacks and indigenous Americans (Provine 2007). Alcohol prohibition, the first major federal drug prohibition, was relatively short-lived, lasting only from 1920 to 1933. Yet, a series of highly consequential experiments with drug control in the United States would follow.

Moral enterprises, such as the development and implementation of drug laws, are undertaken by **moral entrepreneurs** (Becker 1963). Moral entrepreneurs are the powerful people who use their influence to create and administer social and legal rules that define certain groups or behaviors as deviant. In the case of drug prohibition, these include the organizations and individuals who advance relevant laws and bureaucracies and the claim that such rule enforcement serves the public interest and protects moral values. Harry J. Anslinger provides a classic case of a moral entrepreneur (Becker 1963). Previously an employee of the division of the US Treasury that dealt with alcohol prohibition, Anslinger became the first head of the newly formed Federal Bureau of Narcotics in 1930 (Galliher, Keys, and Elsner 1998). As head of the sole agency addressing narcotics policy, Anslinger was able to "define and legitimize his interpretation of the drug problem, to mobilize legislative initiatives, and to implement an official law enforcement plan of action" (Galliher, Keys, and Elsner 1998:667).

Anslinger worked tirelessly to promote narratives linking narcotic substances such as cannabis to insanity, addiction, violence, crime, and racist stereotypes of people of color while actively suppressing research and information that proved otherwise (Galliher, Keys, and Elsner 1998). Countless novels, films, and magazine articles, many authored or produced by Anslinger himself, strengthened connections in the imagination of the US public between drugs and threats, often depicted in the form of people of color, tearing at the moral and social fabric of white middle-class communities. During the early 20th century, whites were the primary users and traffickers of substances such as cannabis, opiates, and cocaine (Helmer and Vietorisz 1974; Galliher, Keys, and Elsner 1998). Yet, the amplification of ethnic stereotypes and fears about deviant and marginalized groups such as Mexican migrant laborers, black men, jazz musicians, artists, and street criminals helped motivate the passage of laws such as the federal prohibition of cannabis in the 1930s via the Marihuana Tax Act (Morgan 1981; Cohen 2006; Provine 2007).

While each case is unique, a general trend between attitudes and laws addressing substances and their association with racial or ethnic groups is evident in much of US history. Stanley Cohen (1972:9), a sociologist, theorized that a **moral panic** takes place when a person, activity, or event becomes "defined as a threat to societal values and interests." Moral panics are often an exaggeration of the actual threat posed. As Stuart Hall, a cultural theorist, and his colleagues (1978:52) argued, institutions like the media and government "do not simply respond to 'moral panics'" but actually "form part of the circle out of which 'moral panics' develop," often by amplifying them. For instance, prominent members of the American media and politicians played a significant role in spreading narratives that stoked fears about Chinese and Mexican immigrants as foreign "aliens" and harbingers of dangerous and exotic drugs.

A racialized moral panic also accompanied the discovery of the addiction potential of cocaine during the Reconstruction and Jim Crow periods in the American South from the early 1910s to the 1940s. During this time, wealthy whites saw themselves as legitimate medical or recreational users of cocaine for everything from parties to toothaches (Cohen 2006). However, politicians and newspapers demonized the black cocaine user as the "Negro cocaine fiend" made insane and violent by the drug (Cohen 2006; Provine 2007). A **folk devil**, or a group of people who come to embody the assumed problem in the public imagination, is a regular feature of moral panics (Cohen 1972). The folk devil of the so-called "Negro cocaine fiend" aligned with dominant racial stereotypes about black men as "brutes" that rationalized centuries of violence, suspicion, and discrimination (Provine 2007).

Examining drug policies in the United States reveals a new perspective on the relationship between laws and society. Legal rules are not just products of the lawmakers' knowledge and engagement with available facts. The passage and enforcement of laws are shaped by the social groups that yield influence over decision-making and the crafting of public narratives.

For moral entrepreneurs, the passage of laws can even depend on the dismissal or minimization of clear and compelling evidence. In 1944, the New York Academy of Medicine released a report about cannabis use and its effects based on extensive research. A group of experts and specialists in the Academy's Mayor's Committee on Marihuana (1944) drew on sociological, physiological, and psychological evidence about cannabis and its use. The study debunked claims that cannabis use caused sex crimes, violence, insanity, or the use of other drugs. Yet, policymakers either ignored or contested the report.

This trend was echoed decades later with the release of the 1972 report of the National Commission on Marihuana and Drug Abuse. The report compiled evidence on cannabis use to guide future drug policies. The Commission found that it was not an imminent danger to society and called for further research. They, therefore, suggested reconsidering cannabis's status as an illegal narcotic and addressing the problem of drug misuse through education and treatment. Regardless, President Nixon, who had commissioned the report, disregarded its findings and recommendations in his positions on drug policy.

Drug prohibition policies and practices helped white elites maintain the disenfranchisement of black citizens from inclusion and influence in the United States in the aftermath of slavery (Provine 2007; Alexander 2012). The prison system in the United States helped maintain the economic, social, and political dominance of upper-class white citizens over other populations (Wacquant 2001; Alexander 2012). For instance, in 1974, the Committee on the Judiciary headed by Mississippi Senator James O. Eastland held hearings in response to the 1972 report on cannabis use. The Committee advocated for continued prohibition and asserted that cannabis use was indeed a dangerous epidemic. That Eastland was a moral entrepreneur in the War on Drugs was not a coincidence. Throughout his legislative career, Senator Eastland had been vocal in his support for racial segregation and his belief in white supremacy in the congressional record (United States Congress 1944).

The connection between drug laws and racial oppression has become even more powerful and institutionalized. Unlike the short-lived nature of alcohol prohibition, prohibitions against substances such as cocaine, heroin, and cannabis have become fixtures of US legal, political, economic, and social systems. The outcomes of these systems continue to disadvantage already marginalized racial and class groups. The rise in imprisonment rates and shifting demographics of US prisons since the 1950s from primarily white to black was a byproduct of the War on Drugs (Wacquant 2001; Provine 2007). Intensified incarceration policies reflected a backlash, in the 1960s and 1970s, against movements for equality and social progress such as the Civil Rights Movement (Haney López 2007; Provine 2007; Alexander 2012).

Throughout the 1980s and 1990s, the prison population and penalties linked to drug crimes expanded (Provine 2007). Police increased their use of military tactics to enforce drug laws (Alexander 2012). Lawmakers also passed laws such as mandatory minimum sentences for drug crimes, sentencing disparities for crack and powder cocaine which disproportionately affected poor and black citizens (Haney López 2007; Provine 2007; Welch 2007). And influential groups amplified moral panics around issues such as the "crack epidemic."

The US' drug enforcement policies and practices continue to punish black people for drug crimes disproportionately compared to other racial groups. Researchers in sociology and criminology have found strong evidence of racial discrimination at almost every step of the policing, judicial, and criminal justice process. This evidence accounts for more of the racial inequalities in drug arrests and imprisonment than other potential explanations, such as the demographics of drug users (Lurigio and Loose 2008). The racial demographics of drug producers and distributors also fail to explain these disparities (Ojmarrh and Caudy 2013). For example, the sociologists and criminologists Katherine Beckett, Kris Nyrop, and Lori Pfingst (2006:121) found, "although a majority of drug transactions involving the five serious drugs under consideration here involve a white drug dealer, 64 percent of those arrested for drug delivery in Seattle from January 1999 to April 2001 were black."

Local police practices of drug law enforcement vary. In San Francisco, a place touted for its progressive approach to drug use, police departments may target predominantly poor and black neighborhoods in response to pressure from local economic elites to gentrify these neighborhoods and attract wealthier, whiter residents (Lynch et al. 2013). In others, such as Colorado, policing may be influenced by cultural depictions of local drug problems (Beckett et al. 2005). Regardless, they conform to an apparent national pattern. Drug policy enforcement targets predominantly black communities more than other populations.

Racial inequality in outcomes within the legal system grows even further at the stages that proceed arrests such as pretrial evaluations and sentencing. Regardless of the nature of the crime or criminal history, across the US, in comparison to black men, white men charged with drug felonies are more likely to receive pretrial diversion and therefore face probation rather than prison (Schlesinger 2013). In California, researchers found that regardless of criminal record or charge, whites charged with drug crimes were more likely to receive drug treatment while blacks were more likely to receive prison (Nicosia, MacDonald, and Arkes 2013). Overall, in comparison to blacks, whites receive shorter prison sentences for the same drug crimes in the United States (Ward, Hartley, and Tillyer 2016). Prosecutors are more likely to suggest lower sentences for whites than blacks (Ward, Hartley, and Tillyer 2016).

The consequences of these forms of discrimination reverberate beyond the judicial and criminal justice system. They impact people's lives, well-being, families, and communities. In her research on the impact of felony convictions on employment, Devah Pager (2007:145), a sociologist, found that "the status of 'ex-offender' is formalized and legitimated by the imposition and dissemination of a criminal record, which are in turn used by employers and other gatekeepers in ways that restrict access to valuable social resources." Using a type of social experiment known as audit study, Pager sent young men who were identical in almost every way aside from their race and whether they identified as having a criminal record out to apply for the same jobs. Pager found that racial discrimination has a long-term impact on the employment prospects of blacks and whites, with white applicants who had felony convictions receiving more callbacks than blacks who did not have a criminal record. And these biases and practices affect people's quality of life in multiple ways. The aggressive policing and incarceration practices linked to the War on Drugs in predominantly black communities also have negative impacts on the health and education of residents (Sewell and Jefferson 2016; Britton 2019).

The War on Drugs Debate

Since the term has grown in its use and popularity, it has become increasingly common to hear the War on Drugs described as a problem. Television shows like *The Wire* highlight the downsides of US drug policy. Popular books like legal scholar Michelle Alexander's (2012) *The New Jim Crow*, neuroscientist Carl Hart's (2013a) *High Price*, journalist Gary Webb's (1999) *Dark Alliance*, and activist Jack Herer's (1998) *The Emperor Wears No Clothes* demonstrate public interest in critiques of the War on Drugs. Additionally, a growing consensus among researchers and scholars from a wide range of disciplines concludes that the War on Drugs has perpetuated racial oppression and other social injustices and caused massive harm to society as a whole. For example, in 2021, a group of social and medical scientists and bioethicists published a research-based consensus statement in the *American Journal of Bioethics* calling for an end to the War on Drugs on these grounds.

At the same time, moral entrepreneurs continue to perpetuate myths and racialized moral panics about drug use and drug policy. In 2019, novelist and media commentator Alex Berensen published a book entitled *Tell Your Children: The Truth About Marijuana, Mental Illness, and Violence*. The book employs horrific anecdotes of acts of brutal violence and biased presentations of data in an attempt to argue that cannabis use causes mental illness and violence. Furthermore, the book claims these presumed issues of cannabis-related psychosis and violent behavior are particularly problematic within black communities. However, despite such attempts at revamping the infamous "Reefer Madness" campaign of the past for contemporary audiences, these arguments and claims have not had the same success in shaping public narratives as Anslinger's earlier efforts.

Upon the book's publication, experts and commentators immediately dismissed and condemned Berensen's claims. Joining the chorus of a multitude of negative reviews, journalist German Lopez (2019) wrote that the book was "essentially an exercise in cherry-picking data and presenting correlation as causation." Moreover, 100 researchers, scholars, and clinical and medical practitioners signed and published an open letter debunking its claims as dangerous "junk science" (Drug Policy Alliance 2019). The letter argued that Berensen distorted and misrepresented the scientific evidence "to uphold and perpetuate the worst myths about people of color and people with mental illness" (Drug Policy Alliance 2019).

So, how did we get from a series of moral panics about drugs, crime, and race and ethnicity that motivated and created the War on Drugs to increasingly thinking about the War on Drugs itself as a **social problem,** an issue recognized as harmful by sectors of our society?

Herbert Blumer (1971:298), a sociologist, argued that "social problems are fundamentally products of a process of collective definition." Debates that take place around kitchen tables and water coolers, in the pages of the *New York Times*, in the halls of Congress, and on Facebook, Twitter, and the comment sections of websites play a significant role in this process. Through media, public debates provide engagement and social information for people on which issues are worthy of consideration, how they should think about them, why the issues matter, what individuals and society should do about them, and the effects of current efforts to address them. This process of collective definition and its relationship to how Americans talk and think about the War on Drugs is a primary focus of this book. I examine the role of racial identities and inequalities and trends in politics and media in the content and context of the War on Drugs debate.

A core component of understanding any issue from a sociological perspective is cultivating and employing your **sociological imagination** (Mills 1959). The sociological imagination is a way of thinking that enables us to connect broader trends and issues in history and society, such as laws and policies, significant events, disparities in access to resources, or even the growth of the internet, to people's everyday lives and experiences. A sociological approach to the War on Drugs debate can tell us about the relationship between peoples' sense of identity and essential issues in society. Within these discussions, people take stances on contentious issues and articulate ideas about themselves, other people, and society.

In other words, the War on Drugs debate can tell us about the relationship between society and identities. **Identity** is something that almost everyone has, but many of us take for granted – yet many of us could very easily tell someone, even a stranger, who "we" are. So, in that sense, we can say that we produce our identities through language or more broadly communication (Howard 2000). Identity also changes over time; it is an ongoing process rather than something fixed (Hall 1996). Just think about how differently you described or thought of yourself when you were much younger in comparison to now (for some of us, this might even be a bit embarrassing).

Identity is also related to social groups. People define themselves by what groups they belong to and what those groups mean to them (Howard 2000). When asked to describe yourself, you might mention something like your nationality, a club you belong to, your profession, or how you classify yourself in terms of social categories like ethnicity, class, or gender. These categories all imply group membership, how vital membership in that group is to you, and maybe even to which groups you don't belong. Pronouns like "we" and "us" suggest how our identities are tied to groups, while "they" and "them" might suggest which groups we see as outsiders or separate from ourselves.

Finally, identity, to use some sociological jargon, is a result of both **agency** and **structure** (Giddens 1993). People often develop and express their sense of who they are strategically by making choices or exercising agency (Goffman 1967). But they don't get to choose the raw materials and opportunities (e.g., ideas, experiences, categories, or even recognition by others) that they have available to build an identity. People hold varying degrees of power over the categories and meanings that institutions and other people use to label and identify them (Bourdieu 1991; Collins 2009; Lewis 2003a). These are, instead, often a product of the way society is organized or, in other words, the social structure. Throughout US history, arguments over racial categories (e.g., whether someone is considered "white" or "black" and what that means) have often been decided and thus made "legitimate" by legal decisions on behalf of the government (Haney López 2006; Omi and Winant 2014).

On the one hand, identity is an almost intuitive and obvious aspect of our everyday experiences; on the other, it is incredibly complex and shaped by large and powerful social forces (Berger and Luckmann 1966). Some questions that help us think about identity from a sociological point of view are the following:

- What are the stories that a person tells or the claims that they make in social life?
- Who are the human characters in those stories and claims?
- How does a person define the "I" or "we" or "us" or "me" in those stories and claims as opposed to the "them" or "they"?
- How does someone's sense of self relate to social categories and social groups?
- How does a person define themselves in contrast or comparison to others?

Understanding all this not only helps us better address social problems connected to issues like the War on Drugs; it also helps us cultivate **sociological mindfulness** (see Schwalbe 2005). In other words, making these kinds of connections helps us comprehend how things like the level and character of inequality in our society shape the situations we encounter in everyday life. But this mindset moves beyond just using our sociological imaginations. Sociological mindfulness helps us reflect upon how what we do in everyday life impacts other people and society in ways that either contribute to or help dismantle social outcomes like inequality (Schwalbe 2005).

When institutions and social systems distribute resources and rewards towards certain groups of people and away from others and influential people construct and enforce rules to the advantage of particular groups rather than others, these arrangements perpetuate **inequality**. We can refer to this ongoing process that maintains inequality as **oppression**. To fully understand unequal outcomes in things like property, prestige, or political power, we need to look at a wide range of historical and ongoing collective actions, institutions, and systems. Unequal outcomes between social groups are not inevitable or natural. They are not merely a product of individual choices, biological inheritance, or innate differences in effort or intellect between individuals. They are a product of the things that people do and have done in the past, in collective and organized ways, to create social divisions and group boundaries that benefit some social groups and disadvantage others.

A significant form of inequality that impacts this public debate, and the War on Drugs itself, is disparities in influence, assets, and opportunities between racial groups in the United States. A mountain of social science research demonstrates that **racism** and **racial inequality**, predominantly advantaging those who are considered white, remains rampant in the United States. Despite significant social and political struggles for equality, racially unequal distributions in both material resources like income and wealth and social and emotional rewards like respect and empathy from others in everyday life persist (Essed 1991; Oliver and Shapiro 1995; Feagin 2006; Bonilla-Silva 2014). Sociologists often describe this state of affairs as the product of systemic racism (Feagin 2006) or evidence of a racialized social system (Bonilla-Silva 2014). But all this talk of "systems" doesn't mean that some abstract nonhuman system causes racial inequality. Racial inequality is a result of the things people do, think, and say that maintain **racial oppression**. And these actions, thoughts, and words have essential patterns and relationships that sociologists strive to understand.

With all this in mind, how people talk and think about the War on Drugs has profound sociological implications. Throughout this book, I will explore four aspects of the debate over the War on Drugs: (1) the history of the relationship between racism and drug policies, (2) the role of the media as a place where people debate these policies, (3) how the debate reflects popular ideas about race, crime, and politics and even commonly held ideals like justice, equality, and freedom, and (4) how people construct and reinforce identities through their participation in these debates and what that means for society.

Understanding the Debate through Research and Data

There is another important aspect of approaching this topic sociologically. Sociology doesn't just involve sitting around and thinking about an issue or problem. Instead, sociologists actively conduct research to try to find answers. **Sociological research** consists of the systematic collection and analysis of data to answer questions. Collecting and analyzing data provides a window to patterns and processes in society into which we wouldn't otherwise be able to peer.

One place to look to try to understand how Americans talk about something in the media is media content itself. This type of research, looking for patterns and themes in the actual content of media communication, is known as **content analysis** (Altheide and Schnieder 2013). I conducted a content analysis of over 30 years of US newspaper content that focuses on the War on Drugs, including 394 op-eds, letters to the editor, and news articles. The patterns that I found demonstrate the social and cultural contours of this heated debate. My analysis of these newspaper manuscripts allows me to show how proponents and critics of the controversial social issues of drug policy and incarceration "frame" their arguments in mass media. That is, it tells us what aspects of the War on Drugs people tend to emphasize.

I analyzed 3,145 comments on the internet to examine the US debate on the War on Drugs. I collected them from the comments sections of online news articles that discuss various aspects of the War on Drugs. Conducting this research meant reading thousands of Internet comments (which I would not personally recommend unless you can write a book about it). Comment sections may seem like an odd or irrelevant source of data. Analyzing these online comments, however, reveals two important things. First, it helps provide an idea of how people react to different arguments and stories about the War on Drugs and how they interact with one another in these discussions. Second, it can tell us more about the reasoning and ideas that members of the public use to discuss this issue and, especially, the arguments and claims that are out there but might not end up in a newspaper.

Through a discussion of the findings and their implications, I show how ideas about racial inequality, politics, society, and crime, and forms of evidence and statistics such as rates of arrest and incarceration or the financial costs of drug policies and incarceration are advanced, interpreted, and contested. Moreover, I show how people form a sense of racial or political identity in debates over policy issues tied to racial inequality such as the War on Drugs through narratives that connect racial categories to concepts such as innocence, criminality, free will, and fairness. Along with the data and findings, throughout this book, I offer a variety of ideas, events, and theories – which appear in bold – that you can use to make sense of these vital issues in contemporary society.

Overview of This Book

Chapter 2: "The War on Drugs as a Contested Social Issue" provides an overview of the role of debates and contestations over social issues. Issues of race, crime, drug use, and public policy are not merely aspects of American society but objects of ongoing debate in the **public sphere**. I discuss first what is meant by the term "public sphere" as both an ideal and as a reality. I then demonstrate how debates over race, crime, immigration, drug use, and public policy function in society. I show how they intertwine with US history and trends such as urbanization (the development of cities and emergence of "urban life"), social statistics such as crime rates, and uncertainty over the fate of a multiracial America after the outlawing of slavery.

Once having established the historical antecedents, I discuss how subsequent forces shape contemporary debates on these issues, including the current debate over the War on Drugs in society. I examine how the use of new technologies, the militarization of drug enforcement policies, and racial ideologies intersect within this debate. I introduce the concept of **contested social issues** and illuminate how social, cultural, and political forces shape the War on Drugs debate.

Chapter 3: "How Does the Media Frame the Debate?" answers its titular question. Mass media plays an enormous role in how contested social issues are discussed and debated in contemporary society. This chapter first explores some examples of the role of mass media in debates over contested social issues. It also provides an overview of how social scientists understand the relationship between media and society. Some of the key concepts I discuss include **framing** and **agenda-setting**. I also present findings from my analysis of newspaper content from 1983 to 2014. I show the various ways that people argue for or against the War on Drugs and how often each type of argument appeared in the newspapers.

I define each frame and the themes within that frame with quotes from newspaper op-eds, letters to the editor, and reporting. These frames include that the War on Drugs is a failure in terms of its stated goals (**the functionalist frame**); that it is too expensive or a poor use of taxpayer funds (**the fiscal frame**); that it violates shared ideals about equality and liberty (**the freedom and justice frame**); and finally that it disproportionately impacts people of color (**the racial unfairness frame**). After discussing these findings, I pose the question of why specific frames appear more commonly than others in the "War on Drugs" debate in newspapers. And it is this question that animates the findings and discussion presented in subsequent chapters.

Chapter 4: "Debate Dynamics: Racial Silence, Code Words, and Resonance" brings clarity to the sociological puzzle presented by the previous findings and reveals the profound impact of racial meanings on this debate. Historical and contemporary evidence is clear that racial meanings play a significant role in the War on Drugs. I examine how racialized meaning and racial logics are filtered into arguments that Americans employ to debate its legitimacy and how racial ideologies influence how the public perceives issues of crime and drug use. I introduce the term **racial silence**, which refers to an absence or refusal to recognize histories of racism in the creation, invention, and enforcement of moral panics.

The chapter also includes a discussion of the concept of **resonance**: a meaningful relationship between cultural patterns, audiences, and cultural products such as media content. Why do certain cultural products resonate (or not) with individuals and groups, and what can it tell us about why specific arguments are more common? I also discuss some relevant trends that might influence the resonance of certain frames in contemporary society. These trends include the rise of colorblind racial ideology (Bonilla-Silva 2014), the prevalence of ideas such as individualism and personal responsibility, the increasing sense that America is a "post-racial" nation, and the changing role of racial language in politics. I show how even though there was **racial silence**, racial insinuations, or **code words** that suggest racial tropes or stereotypes were commonly used by participants in the debate even in arguments that don't overly mention race or racial groups. Ultimately, I explore why there has been a large-scale silencing of arguments that discuss racial injustice or racism in debates within major print media outlets and how this silence places severe limitations on our ability to respond to this issue as a society adequately.

Chapter 5: "Identities Constructed in the Heat of Debate" focuses on one of the most unique contributions of this book by showing how, in debates over contested issues, people are constructing identities and making claims about others. New forms of technology are employed by individuals to assert their sense of self, express their reactions, and actively shape the debates. This chapter focuses more closely on an analysis of online comments on news stories about the War on Drugs. The comments often employ similar interpretative frames that appeared in print media such as newspapers, but unique patterns emerged. So, I also examine how they differ in significant ways. Furthermore, I demonstrate how the Internet comments involve **racial identity construction:** how people produce a sense of their "racial self" and "racial others."

I relate these findings to sociological understandings of race and identity. I uncover how **racial identity construction** takes place by exploring concepts such as **subject-positions** and **symbolic boundaries** (Hall 1997; Lamont 2000; Hughey 2012). I show in this chapter how people produce their racial identity or their sense of racial self through talking about themselves and others in racial terms and the role of conflicts over what racial identities mean. For instance, a sizeable portion of online comments made claims about what it means to be black or white. Commenters routinely related these racial categories to characteristics like morality, criminality, innocence, or victimization.

In **Chapter 6: Conclusion**, I discuss the implications that the debate over the War on Drugs has had for US society and what it can tell us about the relationship between the media, individuals' sense of themselves, public policies, and racial inequality. I reveal the role of **dominant racial meanings** in debates over contested social issues like the War on Drugs and how they diminish our ability to recognize and address the causes and consequences of racial oppression. I advance sociological understandings of the role of racial meanings in the maintenance of racial oppression by examining the connection between racial ideology and racial identity.

Alongside discussing this problematic cycle, I also highlight pathways forward, such as challenging **commonsense myths**, rethinking identity, empathy, and morality, altering our engagement with media, debates, and public policy, and engaging in organized and **collective action**. I also share some organizations that are already doing the work to dismantle the oppressive cycle produced by the relationship between racial inequality, the War on Drugs, politics, and the media.

Discussion Questions

1 What issues come to mind when you think of the War on Drugs? How do these issues affect your life?
2 Make a list of social groups (age, race, ethnicity, religious faith, leisure interests) to which you belong. In what situations do these various identities become more or less prominent in your interactions? Which of these group identities did you choose? Which are involuntary? Which of these identities position you as more vulnerable to discrimination or police violence? How do you negotiate these in your everyday life?
3 What is the difference between a social problem and an individual problem? What is the relationship between these two concepts?
4 Use your sociological imagination to think about something that happened to you today. How does that event reflect broader patterns of power and inequality?
5 In what ways do race and racism structure the forms of violence that Americans of diverse racial and ethnic backgrounds encounter in the United States? Explain.

Chapter 2
The War on Drugs as a Contested Social Issue

Analyzing the War on Drugs enables researchers to understand public problems such as racism, privacy, legislation, police violence, drug use, and mass incarceration. Scholarship across several academic and research fields substantiates that the War on Drugs plays a significant role in maintaining racial inequality by disproportionately targeting, controlling, and punishing marginalized racial groups.[1] Yet, among the broader public, the War on Drugs is a **contested social issue** or an issue of profound importance to our society under contestation, controversy, or debate. These are issues that animate heated discussion or even arguments.

Such discussions provide opportunities for Americans to express their racial identities and worldviews. People not only discuss these issues in the privacy of households or the halls of government but in the public sphere at large through a variety of venues and media. In this chapter, I examine the following: (1) contestation over the War on Drugs; (2) historical debates over race, crime, and immigration; (3) the role of the mass media; (4) militarization of law enforcement; and (5) racism and racialized practices and policies.

The Public Sphere and Media Consolidation

The idea of the **public sphere** comes from the work of German sociologist and philosopher Jurgen Habermas. Habermas argued:

> A portion of the public sphere comes into being in every conversation in which private individuals assemble to form a public body. They then behave neither like business or professional

people transacting private affairs, nor like members of a constitutional order subject to the legal constraints of a state bureaucracy. Citizens behave as a public body when they confer in an unrestricted fashion-that is, with the guarantee of freedom of assembly and association and the freedom to express and publish their opinions-about matters of general interest.
(1964[1974]:49)

In other words, having a venue outside the private domain of households and the space of government, where people can discuss important issues and come to conclusions before engaging in political actions, is vital for a healthy democracy. The concept of the public sphere has been one of the most influential ideas in recent history.[2] Habermas argued that an ideal public sphere involved people coming together and forming a consensus about the best course of action for policies or social change based on rational arguments. He proposed that people can find agreement by "bracketing" (or leaving outside of the public sphere) their allegiance to smaller group memberships to seek the common good of society (Habermas 1962[1989]).

The idea of the public sphere that Habermas proposed represents an idealized space which, in practice, often falls short. Corporate interests and white elites, overrepresented in corporations and government arenas, continue to dominate the public sphere in the United States. Even the historical ideal that Habermas (1962[1989]) analyzed – 18th-century European liberal democratic societies – prohibited large sections of the public based on gender, ethnicity, and class from accessing the spaces and associations where significant debates and discussions took place (Fraser 1992). However, their participants presented themselves as a universal class of citizens – the public.

In real life, the public sphere is fraught with limitations and inequalities. Nancy Fraser, a political theorist and feminist philosopher, reminds us that "despite the rhetoric about publicity and accessibility, the official public sphere rested on, indeed was importantly constituted by, a number of exclusions" (1992:113). Historically marginalized and excluded groups developed parallel public spaces to generate alternative routes for political influence. Women's groups, for instance, met in the private sphere but also discussed gender subordination and related forms of inequality (Fraser 1992). These spaces where marginalized groups can have influence and generate forms of constrained agency inspire social justice movements such as the Civil Rights Movement (Morris 1984). In short, the reality of public discussions and debates is more complicated than the ideal of the public sphere.

Mass media is an umbrella term for the ways that corporate entities distribute information, including television, film, radio, and print and digital media such as blogs, magazines, literature, and newspapers. Mass media provide essential platforms in the contemporary public sphere. In the age of digital and social media, some Americans may perceive that barriers to the public sphere have melted away. In this view, especially in the current era of social media platforms such as *Facebook, Twitter*, and the blogosphere, racialized minorities, women, immigrants, and other marginalized groups have created a parallel public sphere. However, the platforms where debates over contested social issues occur are still profoundly influenced by large-scale political, economic, and cultural trends and forms of exclusion and inequality.

For example, **media consolidation** is a process in which fewer and fewer companies produce media content as media corporations merge and smaller producers are purchased (Bagdikian 1983; Vizcarrondo 2013; Steiner 2015). In 1983, the journalist and professor Ben H. Bagdikian published *The Media Monopoly*, bringing this emerging issue to the American public's attention. He pointed out that only about 50 large conglomerates owned and controlled the majority of media outlets and venues and expressed concern over the impacts of media consolidation on journalism and news reporting. At the time, critics, particularly those allied with media corporations, dismissed Bagdikian's claims as overly alarmist (Hacker 1983). However, scholars and commentators now hail the book as prophetic and influential.

Evidence has only compiled that media consolidation remains a major trend with troubling social implications. In 2000, the political scientist and Senator Paul Wellstone (2000:552) argued:

> For our democracy to work, we depend on the media to do two things. We depend on them to provide citizens with access to a wide and diverse range of opinions, analyses, and perspectives. And we depend on the media to hold concentrated power-whether public or private power-accountable to the people. The greater the diversity of ownership and control, the better they will be able to perform those functions. On the other

hand, as ownership and control of the media becomes concentrated in the hands of fewer and fewer people, the less we can rely upon the media to fulfill these basic responsibilities.

In the years since, this trend has only compounded. In 2019, five multibillion dollar corporations AT&T, Comcast, The Fox Corporation, The Walt Disney Company, and ViacomCBS own the majority (90%) of major media outlets (Lutz 2012; VanDerWerff 2019). With such vast resources at their disposal, they also hold political power to push for further deregulation of the media industry. Even small-scale media producers like local television networks are not immune. Sinclair Broadcast Group, as of August 2017, owned 285 local stations, almost 60% of the market, with its content reaching over 180 million people (RabbitEars 2017). Importantly, ownership influences content. When Sinclair acquires local stations, their broadcasters become more likely to use politically conservative framing and language in discussions of current events and issues (Martin and McCrain 2019).

While the media provides a channel for mass communication and public debate, extremely powerful corporations operating in the economic context of advanced and globalized capitalism produce mass media content. In other words, mass media content producers are motivated by goals that contradict mass media's potential role as a venue for the public sphere in democratic societies (Habermas 1962[1989]; Robins 1997). As the sociologist and media scholar Kevin Robins (1997:33) wrote, "driven now by the logic of profit and competition, the overriding objective of the new media companies is to get their product to the largest number of consumers."

A byproduct of media consolidation is that increasingly small and socially homogenous pool of people who possess the power to shape hold influence over the stories, ideas, and images that appear in mass media (Turner and Cooper 2007). As of 2006, racial minorities owned only 3% of commercial television stations. Blacks and Latinx people owned fewer than 1% (Turner 2007). Because mass media corporations are predominantly owned and operated by white elites and staffed by upper-middle and upper-class whites, their content rarely reflects the concerns of the poor, lower-middle class, or racial and ethnic minorities (Robinson 2000; Entman and Rojecki 2001; Downing and Husband 2005). And the digital revolution has not done much to alter this trend. Online content producers and consumers, much like the print media, remain highly racially segregated (Daniels 2013).

Alongside these more recent trends, we need some historical perspective. An analysis of earlier debates is required to understand contemporary debates. The current discussion on the War on Drugs developed over a century in which the issues of race, crime, and immigration gripped the nation and held dire consequences.

Debates over Race, Crime, and Immigration

Think about how common it is to hear about "urban" or "inner-city" crime. Urbanization, the development and growth of major cities since the early 20th century, has a profound influence on how Americans discuss the issue of crime. In the early 1940s, criminologists began theorizing that urbanization brought **social disorganization** (Shaw and McKay 1942). Socially and economically diverse groups of people live in closer proximity in urban areas. Urban residents are more likely to encounter strangers. So, from this perspective, competition, self-interest, and self-expression in urban life reduce informal forms of social control such as social bonds or shared ideas about what is normal or acceptable (Clinard and Meier 2001). Criminologists argued that urban spaces facilitated criminal acts and created a need for increased formal means of social control such as policing and imprisonment (Shaw and McKay 1942; Clinard and Meier 2001). While these ideas influence how many people think about crime, they leave much to be desired. For instance, how do we explain which groups end up targeted for formal social control and what makes certain spaces any more disorganized than others?

Using in-depth ethnographic research, Martín Sánchez-Jankowski (2008), a sociologist, found that despite their neighborhoods often being described as disorganized, the residents of poor urban communities exhibit high levels of social cohesion and resilience in the face of social change. Additionally, when communities face formal social control measures like incarceration and policing, these conditions exacerbate issues labeled as "social disorganization" such as family disruption and financial instability (Rose and Clear 1998; Clear 2007). Additionally, we cannot understand the causes and consequences of crime without also looking at **criminalization** – the process whereby institutions and authority figures label individuals as criminals (Rios 2011). So, while the "social disorganization" approach might provide tempting explanations, it's essential to maintain what sociologist Stephen J. Pfohl (1985:331) calls a "power-reflexive understanding" of social control measures like the criminal justice system. Issues of power and inequality influence systems and practices of formal social control.

Alongside urbanization and the rise of urban policing, the birth of social statistics and the development of the social sciences also influenced debates about crime. In the early 20th century, with the development of research methods, policing, and advances in mathematics and technology, academics and government officials began to collect, with greater accuracy than ever before, statistical information about social groups (Muhammad 2011). However, these new analytical methods provided limited information. They also did not offer sophisticated concepts and ideas to interpret or even explain these differences. Instead, **scientific racism**, a set of ideas employing concepts and practices from science toward racist ends such as upholding systems of racial oppression, held dominance among white social and intellectual elites (Omi and Winant 1994; Feagin 2010). These ideas, such as that innate or genetic differences between racial groups explained or justified social inequality, have now been thoroughly debunked by social and biological scientists (Graves 2001).

As new sources of information arose about rates of poverty, crime, or unemployment between different racial groups, debates about these issues took on a unique characteristic. What was then under discussion was the meaning and interpretation of such statistical measures as crime and, specifically, how people explained racial disparities in crime rates (Muhammad 2011). There is, of course, a difference between correlation (an observed relationship between two variables) and causation (a relationship in which a change in one variable necessarily causes a change in another).

William Edward Burghardt Du Bois, a sociologist, historian, and one of the most influential public intellectuals, was the first American scholar to systematically research crime (Rabaka 2010; Morris 2015). However, because the United States has marginalized the contributions of black academics, he has often gone uncredited as a formative criminologist. Despite his contributions, including the earliest American sociological research to employ highly rigorous empirical methods and the development of the Atlanta Laboratory, arguably the first American school of sociology, Du Bois was never hired in a tenure-track position at any of the historically white universities (Wright 2002; Morris 2015).

In the late 19th and early 20th centuries, Du Bois collected and analyzed vast amounts of data on urban crime and its social context and developed rich and compelling explanations for crime rates (Rabaka 2010; Muhammad 2011). This research primarily drew on US Census Bureau data and thousands of interviews. Du Bois demonstrated that racial differences in crime rates were a product of residential segregation, disproportionate policing and surveillance, the impact of slavery, racial discrimination, lack of economic opportunities, and lack of government investment in black communities (Du Bois 1889; Gabbidon 2007). He developed nuanced and sociologically informed explanations that conflicted with dominant beliefs. For instance, his interpretations and conclusions clashed with the popular notion that supposedly innate or biological differences between racialized groups explained racial group differences in crime rates (Muhammad 2011). Advocates for racial progress and equality, such as Du Bois, received criticism and censorship from elites and intellectuals who proposed explanations for crime rates that justified racially unequal treatment or skepticism toward the full humanity of blacks (Muhammad 2011).

The ensuing debates around race and crime became especially heightened because they intertwined with what was then called "the Negro question." In the aftermath of chattel slavery, social statistics became part of the heated and vigorous public debates about whether newly freed blacks could (or should) integrate or assimilate into the white-dominated society (Muhammad 2011). White commentators and elites seized upon national crime statistics, such as those compiled by statistician Frederick L. Hoffman in 1896 that suggested black Americans committed higher rates of crime, as proof of their supposed tendency toward criminality as a so-called "race trait" (Muhammad 2011).

Immigration also factored into debates about the demographics and causes of crime. Consider the following exchange in 1912 between two prominent social scientists, Isaac A. Hourwich and Charles A. Ellwood. Hourwich (1912) pointed out that crime rates had no relationship with immigration rates in the United States, and that other causes such as the state of the economy better explained crime rates. Moreover, he noted that native-born white US citizens had higher rates of crime than foreign-born immigrants.

In response to this research, Ellwood (1912) conceded that these statistical relationships were legitimate, but argued that Hourwich's conclusions were hasty for he had not distinguished between immigrant groups predisposed to crime and those that were not. Implying a further connection between immigration and criminality, Ellwood even speculated that ethnic patterns of migration contributed to crime indirectly. Without any supporting data, he argued that competition with foreign immigrants drove up crime among native-born whites of the lower classes. While it took place over a century ago, we can see echoes of this debate in the current public discussion of immigration and crime in the United States.

The effect of these interpretations of crime statistics was akin to what Patricia Hill Collins (2009), a sociologist, calls **controlling images**. They provided depictions of oppressed groups that associated them with features and characteristics that make their oppression seem like a natural outcome. For instance, while ideas of blacks as naturally submissive and docile had previously rationalized slavery (Patterson 1982), a new stereotype of blacks as criminal and threatening helped justify the social sanctions of Jim Crow that emerged after that, including racial segregation by law, surveillance, and punishment (Muhammad 2011). Similarly, notions that immigrants, especially those hailing from non-European nations such as Mexico or China, were predisposed to crime or violence helped rationalize strict immigration policies and formal exclusion from the labor market, full citizenship rights, and benefits from the state (Glenn 2002; Yang 2004; Fox 2012).

Statistics help us measure and understand the social world around us. But without the proper context and interpretation, statistical information can be used for rationalizing harmful social arrangements. Khalil Gibran Muhammad (2011:34), a historian, pointed out that once statisticians had put out reports stating that blacks were more often arrested for crimes, "such empirical evidence could then justify a range of discriminatory laws, first targeting blacks, then punishing them more harshly than whites." In contrast, these statistics were interpreted by racial progressives such as Du Bois as evidence that black Americans needed greater empathy, support, and resources, akin to the support that European migrants had received (Muhammad 2011). Instead, blacks received the social stigma of being associated with crime (Muhammad

2011; Wacquant 2010). And speculation about whether immigration relates to criminality continues to influence debates about immigration policy in the United States (Brown 2013).

Unfortunately, the ideas that emerged from those debates that linked so-called "race traits" to criminality and helped justify racial oppression continue to remain dominant and influential in the contemporary United States. Immigration holds no significant impact on national crime rates, and immigrants tend to have lower reported crime rates than native-born citizens (Martinez and Lee 2000; Butcher and Peihl 2007). However, public opinion research suggests that the vast majority of Americans think that increased rates of immigration likely leads to increases in the crime rate (Rumbaut 2009).

Moreover, crime rates in the United States have declined overall, and the difference in officially reported crime rates between racial groups has consistently declined (Jackson 2013; Males 2013; Denvir 2015; Roeder, Eisen, and Bowling 2015). The crime rate among blacks, in particular, has seen a meteoric drop since the early 1990s (Jackson 2013; Males 2013). The difference that remains is due to ongoing institutional discrimination, residential segregation, mass incarceration, and the exclusion of blacks from fully participating in the economy (Massey and Denton 1993; Reiman 2001; Sugrue 2005; Clear 2007; Pager 2007; Wacquant 2009; Alexander 2012). And yet, whites consistently overestimate the crime rate among black and Latinx populations in survey responses and hold biases that implicitly connect these groups and their communities with crime and violence (The Sentencing Project 2014).

As Blumer (1971) argued, social problems are the product of "collective definitions." The definitions of social problems are the product of not just consensus but also conflicts between groups holding different perspectives and interests. Conflicts over the causes of contested social issues like the War on Drugs, why they matter, and what course of action they demand shape our collective understandings and responses to these issues. Moreover, the venues of public debates over contested social matters change over time. As we can see with the role of media in the public sphere, increasingly our communication on social issues takes place through technologically advanced forms of mass media (Schudson 2011; Couldry 2012). This context means that certain social groups have more power than others in shaping public debates over contested social issues.

Elites and the Power Dynamics of Media Debates

Just as the issues of their time shaped debates at the turn of the 20th century, trends such as the growth of media technologies influence the current discussions on drug policies. Due to technological and social changes, mass media has become increasingly integrated into daily life. As of April 2017, the average person in the United States spends about 12 hours a day consuming some form of media, including almost four hours watching television, over three hours on a smartphone, and over two hours on a computer or laptop (Statista 2017). Media is central to people's habits, rituals, and social interactions in the United States (Couldry 2012). Nick Couldry (2012), a media scholar, argues that media has "supersaturated" society. So, the media continues to shape public debates and the ways people perceive crime, race, and policing.

The media routinely offers spaces for debate. Debates enable media producers to present events and trends in ways that appear more dramatic. They furnish audiences with a sense that they are getting both sides of the story and therefore becoming more informed (Hall et al. 1978). The image of commentators engaged in a heated debate has become an iconic feature of media. CNN's popular and long-running television series *Crossfire* pitted liberal and conservative pundits in intense debates over the issues of the day. *The New York Times*' ongoing *Room for Debate* segment involves a handful of experts weighing in with short opinion pieces on topics such as drug policy, affirmative action, same-sex marriage, tax rates, and income inequality. And the growth of interactive internet sites such as social media has shifted the barriers to entering mediated debates. It's not uncommon to see friends, relatives, and even strangers debating issues such as climate change, racialized police violence, or gun control in cyberspace.

The contemporary mass media provides a place for commentary on events and issues (Couldry 2012). Journalism in the early 20th century clung to ideals of neutrality and objectivity and took on a focus of reporting facts and information rather than opinion (Hall et al. 1978; Schudson 1978). Yet, in the latter half of the century, media forms, including opinion editorials or op-eds in newspapers and commentary and debate-oriented television programming, opened up new vistas and styles of media communication (Jacobs and Townsley 2011). News media remains a site of journalistic reporting of facts presented as objective or impartial (Schudson 2011). But the media now also features what sociologists Ronald N. Jacobs and Eleanor Townsley (2011) call "the space of opinion" with unique consequences for contestations over social issues in the public sphere.

In earlier centuries, dominant group interests that sought to preserve systems of racial oppression such as slavery or Jim Crow were influential in public debates (Muhammad 2011). These same systems of racial oppression now operate in slightly different forms, such as mass incarceration (Wacquant 2001). Stuart Hall and his colleagues (1978:57) argued that in often indirect ways, the media has served to "reproduce the definitions of the powerful." For instance, to preserve the impression that the news reported in major media outlets is impartial or neutral, news sources tend to rely on powerful people who represent established institutions and organizations in society as information sources (Hall et al. 1978; Schudson 1978). Hall and colleagues referred to those who became information sources for the media as **primary institutional definers** and argued that:

> [...] the structured relationship between the media and the primary institutional definers [...] permits the institutional definers to establish the initial definition or primary interpretation of the topic in question. This interpretation then 'commands the field' in all subsequent treatment and sets the terms of reference within which all further coverage or debate takes place. Arguments against a primary interpretation are forced to insert themselves into its definition of 'what is at issue' – they must begin from this framework of interpretation as their starting-point.
> (Hall et al. 1978:58)

In a related way, Howard Becker (1963) found that which groups can amplify and spread their message through mass media influence how rules become developed and enforced in society. This insight is extremely relevant when we think about all the evidence presented in the previous chapter that drug law enforcement takes place in unequal ways that disproportionately punish black and Latinx people. In explaining why some rules get enforced, and not others and that rules are enforced more severely against some social groups and not others, he argued:

> The problem of rule enforcement becomes complicated when the situation contains several competing groups. Accommodation and compromise are more difficult, because there are more interests to be served, and conflict is more likely to be open and unresolved. Under these circumstances, access to the channels of publicity becomes an important variable, and those whose interest demands that rules not be enforced try to prevent news of infractions.
> (Becker 1963:127)

Interested groups use "the available media of communication to develop a favorable climate of opinion" (Becker 1963:145) to create and enforce rules such as drug policy. So basically, media content matters. The media shapes what story is told, what events are reported as significant, and how those events are presented, which all have profound impacts on the rules of society.

These two perspectives help us see how in the "process of collective definition" (Blumer 1971), some people or groups of people have more influence than others. Examining the relationship between mass media and contested social issues not only helps us better understand how society addresses problems (via rulemaking and enforcement, for instance), but also how social groups become identified as "problems" by association.

The Militarization and Racialization of Drug Laws

Alongside the relationship between media, debates, and society, the War on Drugs itself deserves further attention and context. The War on Drugs reflects two fundamental aspects of US drug control policies and practices. First, drug law enforcement practices are racialized or connected to racial categories and inequalities. Second, these practices are militarized. The strategies used to enforce drug laws and to curb drug use in society tend to center on the use of force, imprisonment, and violence rather than other means.

Despite that, as noted by legal scholar Jamie Fellner (2009:266), "for the last twenty years [...] whites have engaged in drug offenses at rates higher than blacks," this imprisonment, force, and violence are more likely to be experienced by blacks and Latinx (Beckett,

Nyrop, and Pfingst 2006). In short, moral panics and stereotypes continue to overshadow evidence. For instance, recent research shows that perceptions that drug dealers and drug-related criminals are predominantly black and Latino alone can increase support for punitive drug laws among white college students (Garland and Bumphus 2012).

While moral panics about race, drugs, and crime have long histories in the United States, the 1980s was a pivotal time for drugs, policing, and racial inequality. Anti-drug rhetoric in the Nixon-era tended to focus on the abstract enemy of drug addiction. However, the Reagan administration (1981–1989) emphasized the human enemy of drug dealers and drug users themselves, thereby disinvesting in previous public health strategies for addressing drug misuse (Balko 2013).

In his first year in office, President Reagan signed the **1981 Military Cooperation with Law Enforcement Act**, modifying previous laws against domestic military operations and opening the floodgates allowing military weaponry, equipment, training, and intelligence to pour into police departments (Alexander 2012; Balko 2013). Police in major cities all over the country began coordinating with the US military and the Pentagon to acquire military equipment to fight the emerging drug war in America's streets (Alexander 2012; Balko 2013). The US War on Drugs also took on an international component with the militarization of the US-Mexican border and increased overseas military operations and interventions under the justification of stemming drug production and distribution (Paley 2015).

Domestically, the use of SWAT raids to enforce drug warrants at homes, apartment buildings, and even public schools grew exponentially from the 1980s till now, often in situations where the use of force is entirely inappropriate (Alexander 2012). Legal scholar Michelle Alexander (2012:75) provides a harrowing description of the traumatic experience of military-style drug raids: "police blast into people's homes, typically in the middle of the night, throwing grenades, shouting, and pointing guns and rifles at anyone inside, often including small children."

These trends in policing practices gave rise to what Radley Balko, an investigative journalist, called the warrior cop (2013). Legal scholar Mallory Meads (2016:636) noted that even the name War on Drugs itself "blurs the distinction between cop and solider and creates a battlefield mentality." The policing style encouraged by the War on Drugs has increased police brutality against the residents of predominantly black communities (Cooper 2015). The warlike mentality that accompanies the War on Drugs has not just resulted in increased use of force and violence by police, but also "racial profiling, psychological intimidation, harassment of citizens, pretextual stops for trivial infractions, and selective enforcement of the law" (Meads 2016:636).

Throughout US history, the racialization and militarization of law enforcement have been overlapping processes. They have shaped one another as they influenced the rules of authority in social systems. For instance, US law enforcement officers often experience conditioning during their training relating to the identification of threats and the use of force against a hypothetically dangerous adversary (Stoughton 2015). Simultaneously, racial stereotypes that equate blackness with criminality and whiteness with innocence shape the dominant public image of this abstract adversary (Welch 2007). The emergence and continuance of the War on Drugs were not only due to changes in policing tactics and criminal justice policies. It was also a mechanism for white elites to double down on previous ideas about which social groups should be policed and punished.

The racialization and militarization of drug control strategies produce striking double standards. Consider the popular images surrounding the idea of the "drug crime epidemic" in the United States. These images were not coincidental or merely reflective of the demographics of this problem. They were, instead, crafted to shift attention toward marginalized populations. During the Reagan and Bush eras (1981–1993), presidential declarations, organizations such as Partnership for a Drug-Free America, and news media outlets who parroted the narratives of government officials depicted the "drug problem" as solely the domain of impoverished black and Latinx people residing in inner cities (Elwood 1994).

These organizations and politicians remained comparatively mute on the sale and use of drugs among politically conservative adults and suburban and white teenagers (Elwood 1994).[3] Despite a rise in cocaine usage in the 1980s and early 1990s among middle- and upper-class white communities, the immense economic and political power contained within these populations made them a less advantageous target for drug law enforcement (Tonry 1994; Wacquant 2009). While measuring actual rates of criminal behavior is difficult, evidence suggests that whites were more likely than other racial groups to have sold drugs during this time and that this trend continues (Case and Katz 1991; Fairlie 2002; Ingram 2014). However, US history, including the post-Civil Rights era, is packed with instances of backlash from white elites against social and political change seen as threatening to their interests (Steinberg 1995; Hughey 2014; Anderson 2016). Lower-class communities of color were thus declared enemy combatants in the War on Drugs.

In contrast to the image of the "drug problem" advanced by drug warriors, the imagery promoted by mainstream drug policy reform organizations has centered on how drug prohibition impacts the future or promise of middle- and upper-class suburban whites (Provine 2007; Hart 2013b). The "whitening" of the substance's users in the public imagination facilitated policy shifts and public support toward decriminalizing cannabis over the past several decades (Provine 2007). Accordingly, reformers have routinely highlighted the bright futures of white middle-class college students and veterans. In these narratives, reformers have emphasized that these individuals are deserving of empathy, health, and safety, and have much to lose if they face punitive sanctions such as a criminal record or prison time (Provine 2007).

These divergent narratives around drug users and criminals from different racial groups helped perpetuate two distinct tiers in our society's response to illegal drug use: treatment and punishment (see Hart 2017). These tiers influence the policies and everyday activities of our legal and criminal justice systems. In a sense, these tiers also represent different emotional relationships between authorities and the public. The treatment tier, the one more often reserved for upper- and middle-class whites, is characterized by empathy and concern. While at the same time, the punitive tier, the one best described as the War on Drugs, is marked by apathy or even antipathy and retribution. As we will see in the debate itself, these emotional relationships play an essential part in the ways that people interpret the War on Drugs and rationalize its impacts on various social groups.

Lawmakers have codified these tiers into law. The **1994 Violent Crime Control and Law Enforcement Act,** written by Democratic Senator Joseph Biden and signed by President William J. Clinton, amplified the punitive tier of the War on Drugs (Johnson 2014). The Act instituted the "Three Strikes Law" mandating that those convicted of a "serious violent felony" in federal court that have "two or more previous convictions in federal or state courts, at least one of which is a 'serious violent felony' (the other offense may be a serious drug offense)" receive a life sentence (FindLaw 2013). Given existent racial disparities in criminalization, the Act further exacerbated racialized mass incarceration (Johnson 2014). Moreover, while it helped institutionalize the punitive tier, many measures for crime prevention and accountability in policing in the Act that represented an empathetic approach were never fully implemented (Collins 1998; Johnson 2014).

Let's take a moment to consider why the emotional orientations of these tiers matter and how they operate. On the one hand, structural barriers such as laws, sentencing guidelines, and access to legal information and lawyers can have a clear impact on how someone fares in these systems (Reiman 2001). However, the daily decisions made by authorities also matter. As argued by sociologist Philomena Essed (1991:2), racism and racial inequality are "routinely created and reinforced through everyday practices." Racial meanings, or perhaps even racial feelings, shape people's ability to draw empathy from police officers or judges rather than skepticism or condemnation (Taslitz 2013). But it's their job to be impartial or rational and to provide justice. So, how does this happen?

Around certain types of people or in certain situations, we may experience a rush of adrenaline or cortisol (a chemical released in response to stressful events) and feel tense or anxious (Gropper and Smith 2013). In other contexts, we may experience a drop-off in dopamine (a chemical related to pleasure and cognition) and feel highly apathetic (Vergne 2016). In our decision-making and social interactions, these feelings can manifest in distrust, the perception of threat, a lack of empathy, or a desire for avoidance. Though they may involve biological processes, these reactions are not natural or universal. Social arrangements such as racial oppression and patterns in mass culture such as ideas, narratives, and images about racial groups influence how people experience **affect** or emotions, sentiments, and feelings in everyday social settings (Thomas and Brunsma 2013).

Patterns in racialized group representation in the mass media inform how police officers, judges, and other legal and criminal justice practitioners think, feel, and treat other Americans (Welch 2007; Taslitz 2013). For instance, analysis of body camera footage by researchers demonstrates that police are more likely to act politely and respectfully in interactions with whites than blacks during traffic stops (Voigt et al. 2017). And these forms of discrimination abound along the criminal justice pipeline. The racial background of someone arrested for drug possession, controlling for all other relevant factors, predicts their likelihood of receiving addiction counseling or jail time and whether their case will even go to court in the first place (Nicosia, MacDonald, and Arkes 2013; Schlesinger 2013).

Beginning in the mid-2010s, the use of opioids, particularly heroin, and the tragic deaths of white Hollywood celebrities, including Phillip Seymour Hoffman, who have "humanized" drug addiction in the eyes of the public, has generated further demand for a softer approach to drug use (Hari 2015). In contrast to the approach to drug use in urban and poor communities, white politicians and the media are now advocating for investment in mental and medical health care, and resources for rural and urban white drug users. Drug epidemics associated with white users such as heroin or pharmaceutical opiate abuse in suburban and rural areas land on the tier of treatment (Netherland and Hansen 2016). Rather than incarceration, surveillance, or violent policing, the opioid epidemic among whites has met calls for empathy, education, treatment, and harm reduction measures (Netherland and Hansen 2016). Even as opioid-related overdose deaths began to increase for blacks in the first half of the 2010s, whites continued to have a 35 times higher likelihood of accessing the life-saving drug buprenorphine, which reduces fatal overdose risk (Lagisetty et al. 2019).

Yet, the militarization of drug policy enforcement continues despite the emergence of a tier of softer and gentler approaches. Moreover, the militarization of police has not been limited to the War on Drugs, though these trends are intertwined. Militarized police forces brutalized peaceful protestors and members of the Occupy Wall Street movement in 2011. Beginning around 2014, heavily armed and armored police forces that resemble a war-time occupation or military invasion have been the government's primary response to peaceful protests and civilian disobedience in response to the almost weekly police killings of black citizens.

Images of peaceful protesters standing adjacent to SWAT teams and police officers in full riot gear pointing weapons have circulated the global media. These events and images are now so routine that they no longer appear shocking. Such shows of force are not only worrisome but largely ineffective in de-escalating conflict. As military veteran Paul Szoldra (2014) argued, "you can't win a person's heart and mind when you are pointing a rifle at his or her chest."

Racial and political conflict around criminal justice reforms and drug policy has continued into the 21st century. The racialization of policing and incarceration has continued and taken on new forms. Reforms touted as making the process of assessment fairer and more efficient have instead embedded racial biases into seemingly neutral and objective technologies. For instance, the for-profit company Northpointe developed a risk assessment algorithm that assigned a score which was used in courts in several states across the company to determine whether someone should be set free or held in detention at various points in the process once charged with a crime. This assessment tool, and many others like it, has proven to be both extraordinarily unreliable and also biased against black and Latinx defendants (Angwin et al. 2016; Hamilton 2019).

Risk assessment tools used in the court systems often include factors such as stable employment, income, social connectedness, and educational achievement, which reflect social positions and inequalities, as indicators of whether or not someone will commit a crime in the future. An analysis of 7,000 cases from Northpointe in Broward County, Florida, revealed that even when controlling the defendant's charge and previous criminal record, "Black defendants were still 77 percent more likely to be pegged as at higher risk of committing a future violent crime and 45 percent more likely to be predicted to commit a future crime of any kind" (Anguin et al. 2016).

Moreover, support for a shift away from highly punitive drug laws such as mandatory minimum sentences has grown. Policy reforms on this front, advanced by former Attorney General Eric Holder, were achieved during the Obama administration (United States Department of Justice 2015). However, beginning in 2016, the Trump administration sought a full return to punitive and aggressive approaches to law enforcement and to further associate the social problem of drug abuse with imagery of violence and people of color. The 2017 remarks by former Attorney General Jefferson Beauregard Sessions III encapsulate this perspective:

> Those of us who are responsible for promoting public safety cannot sit back while any American community is ravaged by crime and violence at the hands of drug traffickers. We can never yield sovereignty over a single neighborhood, city block, or street corner to drug traffickers.

Also, under President Trump's strong leadership, this country is finally getting serious about securing our Southern border. Most of the heroin, cocaine, methamphetamine, and fentanyl in this country got here across border brought here by powerful Mexican drug cartels.
(United States Department of Justice 2017)

A 2017 memo from the Department of Justice rescinded previous sentencing guidelines to advance a principle that "prosecutors should charge and pursue the most serious readily provable offenses" (Office of the Attorney General 2017:1). While the Obama administration enacted restrictions on some forms of police militarization, the Trump administration reopened the direct sale of large-caliber weaponry and grenade launchers from the US military to the hands of police departments (Lucas 2017). As argued by sociologists Mike Vuolo, Joy Kadowaki, and Brian C. Kelly (2017), Sessions, like Harry J. Anslinger before him, was a moral entrepreneur. During his stint as Attorney General, he used this position to dismiss evidence that contradicted his policy goals, posit a connection between violence and cannabis use, and depict it as a hazardous and addictive substance (Vuolo, Kadowaki, and Kelly 2017).

The War on Drugs contributes to racialized mass incarceration – the immense expansion of the predominantly black prison population in the United States. Sociologist Loïc Wacquant (2010) argues that we should refer to this issue as "hyperincarceration" as rather than penalizing the "mass" of individuals in the United States, it largely penalizes a specific subset of the population, impoverished and working-class black men. Since the 1950s, US drug policies and enforcement practices influenced a boom and a complete racial demographic shift in the prison population (Wacquant 2001).

In short, the War on Drugs contributes to a **racialized system of social control** (Tonry 1994; Provine 2007; Alexander 2012). No doubt, this system operates alongside other systems of oppression and inequality, such as social class. For instance, anyone who is in poverty, regardless of their racial identity, is more likely to experience the adverse effects of the system of incarceration, and the influence of wealthy elites in shaping laws and policies is well documented (Clegg and Usmani 2019). However, focusing on racial oppression and social control does not deny this reality. It helps us see these issues as inseparable. Economic disparities in income, debt, and wealth are also drivers and products of racial discrimination and inequality (Oliver and Shapiro 2019; Seamster 2019). The criminal-legal system of policing, courts, and incarceration maintains racial oppression by labeling and regulating disadvantaged racial groups (Coates 2003a; Rios 2011). But given our previous discussion about media, debates, rules, and power, how do these racial implications impact the public conversation about the War on Drugs in the United States?

In the next chapter, we will look at how mass media "frames" the debate. This analysis will illuminate two issues that build upon earlier chapters. First, what do Americans find problematic about the War on Drugs? Second, why are some arguments privileged over others?

Discussion Questions

1. Think of some examples of depictions of crime and criminals on television? If you had no direct or personal experience with crime or the criminal justice system, what conclusions might you draw about crime in the United States based on these media depictions?
2. Identify an issue that is contested among your peers. Explain why this issue is contested. In other words, why do you think this issue is controversial or a source of discussion and debate?
3. Identify examples in the news, the media, political campaigns, or even on your Facebook or Twitter feed where contested issues are debated. Describe what perspectives are not considered or left out of these debates. What might be some of the consequences of reducing problems to only having "two sides?" What might be some of the effects of assuming that "both sides" of a debate are equally legitimate?
4. How are statistics used in contemporary debates about social issues? Who produces these statistics? What are the limitations of using statistics in making an argument?

Notes

1. The scholarship on this topic is too large to list in full, but the studies listed in the previous chapter that demonstrate discrimination and inequality as well as books and articles that provide overviews of the evidence are good indicators of this near consensus among many experts (see Tonry 1994; Provine 2007; as examples).

2 For instance, because of the importance of this concept, Habermas was declared second in 2013s *Global Thought Leaders* list (Frick, Gloor, and Gürtler 2013). This list seeks to track and rank the people who produced the most influential, noteworthy, and innovative ideas.
3 Elwood (1994:12) points out that the media ignored "First Lady Nancy Reagan's possible addiction to prescription tranquilizers" and "President Bush's involvement in arranging a cocaine purchase across from the White House to obtain a powerful visual aid for his televised drug war declaration" because they did not fit the dominant definition of "the drug problem."

Chapter 3

How the Media "Frames" the Debate

During most of the 20th century, print newspapers and television were the dominant forms of mass media. Television programming routinely included live debates between individuals presented as "experts," including political leaders, academics, and public intellectuals. One of the earliest major televised debates over drug policy took place on the television program *Firing Line* (1991). William F. Buckley, a white Republican conservative TV commentator, debated Charles Rangel, a black Democratic congressman who represented the sixteenth district in New York, about the War on Drugs. Charles F. Buckley was an influential conservative intellectual and founder of the conservative publication *National Review*. For much of his career, Buckley stood in opposition to the civil rights movement and racial equality, arguing that blacks were culturally inferior, and therefore not worthy of full social inclusion, and he gave a platform to similar arguments in his magazine (Schultz 2015). Yet, unlike Buckley, Rangel supported militaristic and punitive drug policies, even encouraging Richard Nixon "to ramp up drug-fighting efforts more aggressively, more rapidly" (Mann 2013) during a private meeting.

Today, opposition to the War on Drugs is commonly associated with politically progressive politicians and citizens.[1] So it may be surprising from the vantage point of our current political culture that Buckley argued against drug prohibition. Yet the arguments that Buckley advanced centered on issues of fiscal responsibility, lack of economic productivity, immoral and violent "drug pushers" enabled by the illicit drug market, and the failure to achieve law and order in society. His arguments opposing punitive drug policies appealed to conservative political principles and resentment toward racial groups stereotypically associated with crime and the drug trade.

I begin this chapter with the debate between William F. Buckley and Charles Rangel, because their exchange demonstrates the role that the media plays in circulating dominant positions by white elites that informed public debates about the War on Drugs as well as shaped policies. Buckley's arguments represent an alignment of racially and politically conservative ideas that continue to influence the claims and arguments of this debate. I focus on two of the most important and useful conceptual tools for understanding power and media: framing and agenda-setting. First, I detail how framing and agenda-setting work and the characteristics of the relationship between media and society. I then examine how the participants in the War on Drugs debate have framed the issue in print media content. Finally, I conclude with how the different frames inform perceptions about racism and racial inequality in the United States, which defines the War on Drugs as a social problem.

Setting the Agenda and Framing the Debate

The media is one of the most powerful and influential social institutions in the 21st century (Couldry 2012). The relationship between media and society has become a significant area of sociological research. Researchers who focus on **agenda-setting** are interested in how media can influence what is deemed as important by the public. In other words, the media sets the agenda of public debate and discussion by concentrating on particular issues and events rather than others. In engaging in agenda-setting, the media filters reality by directing public attention to specific problems, issues, trends, and situations. This approach reveals how the subjects depicted, discussed, and debated in the media can lead people to believe that those issues are of more considerable significance (cf. Rogers and Dearing 1988; McCombs, Shaw, and Weaver 1997; Rogers, Hart, and Dearing 1997).

Researchers of politics and the media have used a wide variety of cases and methods to document and understand agenda-setting. In the last chapter, we looked at sociological theories from the 1960s and the 1970s that explained how powerful groups and individuals influence media content and how that influence shapes the rules of society. And since then, with advances in social science research methods, studies have added even more empirical weight to these theoretical claims. Katherine Beckett (1994:425), a sociologist, demonstrated that while "crime and drug use have received unprecedented levels of political and public attention in recent decades," media coverage of these issues does not correlate with actual trends in the rates of crime or drug use. Instead, Beckett (1994) proved something much more concerning – the state and media producers played an outsized role in collectively defining these issues as having political and public significance.

Framing theory, drawn from Erving Goffman's (1974) essay, focuses on identifying frames or "schemata of interpretation" that organize and present social reality. In the context of mass media, **framing** describes how complex real-world events and issues get translated into coherent narratives presented in the media. This process is like photographic framing (Downing and Husband 2005). Within a specific "frame," some aspects of an event or issue are excluded. Thinking about media content via "frames" provides a more nuanced account of what the media does than agenda-setting. Identifying the frames within media content reveals how the media not only convey what events and issues audiences are significant and worthy of attention, but also how audiences should think about those events and issues (cf. Fairhurst and Star 1996; Deetz, Tracy and Simpson 2000; Smetko and Valkenburg 2000).

If this all seems a bit abstract, thinking of a specific event or issue and its depiction in the media can help clarify the power of frames. Take the controversy and mass media debate around Colin Kapernick, a professional quarterback formerly playing for the *San Francisco 49ers*. In September 2016, Kaepernick began refusing to stand and instead took to his knee during the national anthem before his games to peacefully protest ongoing issues of racial injustice such as police brutality. This event is complicated and tied to other contested social issues. Accordingly, there are multiple ways that the media might interpret or frame this act. Is his protest a matter of free speech and the first amendment? Does it reflect the traditional importance of rituals such as playing the national anthem and symbols like the American flag in the United States? Is it part of a long history of black athletes using their celebrity and influence to call attention to racial injustice in the United States?

While agenda-setting would influence whether Americans perceived Kaepernick's actions as worthy of discussion, how the media frames this event affects how many in the United States think about his actions, including why people believe it is important or worthy of attention. And within the context of a debate between two opposing views, framing is especially consequential. So, whether someone supports or condemns this act of protest, they will frame the event in a way that both expresses their view and attempts to convince others that their opinion is the most compelling one.

In short, framing can influence the ways that people judge an event or issue. For instance, political scientists Jon Hurwitz and Mark Peffley (2005) captured the power of **group-centered frames** through an innovative experimental study. The researchers randomly manipulated a national survey question so that half of the survey takers received an item about "violent crime." In contrast, the other half received a question about "violent inner-city crime" with the term "inner-city" acting as a racialized code word or a phrase implicitly associated with a particular racial group (Hurwitz and Peffley 2005).

Their manipulation of the framing of the issue of violent crime in the questions proved useful. Moreover, the experiment affected different groups of respondents based on their racial attitudes. Those who expressed less support for racial progress and equality exposed to the term "inner-city" were more likely to endorse punitive approaches to crime. And those who expressed more support for racial progress and equality exposed to it were less likely to support these approaches. Hurwitz and Peffley (2005:109) noted that:

> when messages are framed in such a way to reinforce the relationship between a particular policy and a particular group, it becomes far more likely that individuals will evaluate the policy on the basis of their evaluations of the group.

Experiments and statistical studies of media agenda-setting and framing have some limitations for helping us understand processes that produce these outcomes. Research combining qualitative and quantitative approaches, however, allows us to see patterns and variations in the actual content of framed media messages and their meanings. By analyzing a large dataset of discussions of the War on Drugs in print media, I identified frames and various themes within each frame. I also looked at the strategies people used to present a claim as valid or rational. Taking this type of approach helps answer the question of how the media frames the debate over the War on Drugs. It helps us understand what kinds of frames appear in discussions of this issue and the patterns among those frames.

Analyzing the War on Drugs: Debates in Print Media[2]

To understand how mass media has framed arguments about the War on Drugs, I examined newspapers, an essential form of print and digital media. Using a digital research database called *Lexis Nexus*, I acquired digital copies of 394 national, regional, and local US newspaper manuscripts containing the phrase "war on drugs." These manuscripts came from 121 newspapers from across the country and included 173 op-eds, 154 journalistic articles, and 68 letters to the editor.

The database included digitally cataloged newspaper content that fits these criteria going back to 1983 and as recent as 2014. So, this data enables a look at the War on Drugs debate over more than three decades. It includes newspaper coverage of important events and issues such as militaristic drug enforcement and the so-called "crack epidemic" and contemporary styles of thinking and talking about race and racial inequality (Haney López 2007). My analysis of newspaper content revealed the various ways that people argue for or against the War on Drugs and how often each type of argument appeared. Table 3.1 presents an

Table 3.1 Frames and Themes in Newspaper Articles

Frame/Theme	N	%
Frame: Fiscal	**103**	**0.1512**
Bad for economy/industry	10	0.0971
Too expensive	93	0.9029
Frame: Freedom and Justice	**167**	**0.2452**
Social class	16	0.0958
Human rights	25	0.1497
Police militarization	9	0.0539
Corruption/greed	29	0.1737
Mass incarceration/overcrowding	47	0.2814
Civil liberties	41	0.2455
Frame: Functionalism	**352**	**0.5169**
Failed/unwinnable	93	0.2642
Doesn't reduce crime/drugs	125	0.3551
Education	21	0.0597
Treatment	65	0.1847
Regulation	48	0.1364
Frame: Racial Unfairness	**59**	**0.0866**
Motivated by racism	5	0.0847
Cause of racial inequality	10	0.1695
Black community susceptible	8	0.1356
Laws/policing are racially biased	36	0.6102
Total	**681**	**100.00**

overview of the number of times each frame and theme appeared in the sample of newspaper manuscripts.

The Fiscal Frame

The first frame used in newspapers to criticize the War on Drugs is what I call the **fiscal frame**. This frame appeared in about 15% of the total claims critical of the War on Drugs in the dataset. The fiscal frame focuses on the financial aspects of the War on Drugs. These elements include relevant costs, where that money is coming from, and even its effects on the economy. Essentially, framing the War on Drugs in this manner allowed critics of the War on Drugs to make appeals to the ideal of fiscal conservativism.

Almost 90% of the claims using this frame articulated that the War on Drugs is simply too expensive. For instance, one letter to the editor argued, "what happens if we legalize drugs?: First, we save the U.S. taxpayer $70 billion per year. This savings comes from the costs of police, courts and prisons all related to drug use" ("Drug Legalization," *Pittsburgh Post-Gazette*, March 16, 2014). As demonstrated by that quote, these claims often included statistics about how much money gets spent on policing, administration, and imprisonment related to the drug war:

> Wyoming spent approximately $9.1 million enforcing marijuana laws in 2010.
> ("Wyo. Pot Laws Harm Many," Wyoming Tribune-Eagle, February 18, 2014)

> With the cost approaching $1.3 trillion, don't you think it's time we stop shoveling huge piles of cash into this bottomless pit of insanity called the war on drugs.
> ("Our Insane War on Drugs," *Intelligencer New Era*, December 19, 2012)

Writers also emphasized per-person costs. One article noted, "jailing someone in Vermont for a week costs $1,120" ("Class war on drug users," *The Buffalo News*, 2014). Another stated, "the $25,000 per person it costs to incarcerate someone for a year in state prison" ("Speaker at Baltimore's Calvert Institute symposium advocates rethinking the war on drugs," *The Daily Record*, April 8, 2005).

The overall economic implications of the War on Drugs appeared in about 10% of these claims. These implications included the loss of the economic activities and potential tax revenue generated by the legalization and taxation of substances such as cannabis:

> If we taxed marijuana at the same rates as cigarettes and tobacco, we would generate about $40 billion in tax revenue. It's about $110 billion a year.
> ("Newton Says Drug War Failing,"
> *Alamogordo Daily News*, January 15, 2013)

The theme of economic opportunity loss also included claims around the loss of labor and productivity caused by mass imprisonment:

> Making users non-criminals would cut jail inmates by half. In addition to prison cost, we lose the productivity they could provide were they not forcibly removed from the work force.
> ("Contra Costa Times Sunday Forum:
> Does the war on drugs need to change?,"
> *Contra Costa Times*, August 6, 2011)

The Freedom and Justice Frame

The second frame, the **freedom and justice frame,** accounted for about a quarter (24.5%) of the claims made criticizing the War on Drugs. Specifically, this frame included presenting the War on Drugs as a problem because it violated widely held ideas about freedom, equality, liberty, and justice in American society without any explicit reference to racial justice, racial inequality, or liberty or freedom for people of color.

One theme, accounting for about 10% of the total claims within the freedom and justice frame, depicted the War on Drugs as an issue of socioeconomic inequality or class conflict. Some authors stated that attitudes on the War on Drugs reflected political beliefs about the poor. An article argued, "with some exceptions, Republicans remain intent on treating drug users as reprobates, especially if they are poor," and concluded that "the war on drugs is a class war on drug users" ("A Class War that Yields no Winner," *Newsday*, February 9, 2014).

Other articles in this theme attempted to lay out the specifics of a class-based double standard that render the poor more likely to face punishment:

> Of course, the drug-offending children of rich parents were not affected, because they didn't need student aid. They were also less likely to get caught and, if they did, could afford better lawyers. But hundreds of thousands of low- and moderate-income students were denied federal aid, often for being found with a stick of marijuana. Different rules certainly apply at the top of the power pyramid. Avid drug warrior George W. Bush had admitted to smoking pot and refused to deny cocaine use – while assuming none of this should disqualify him from being president.
> ("War on Drugs becomes an issue of class,"
> *Chicago Daily Herald*, February 9, 2014)

A more common theme, which made up almost 15% of this frame, depicted the War on Drugs as problematic due to the idea of *protecting human rights*. Articles argued that the policies and practices of the War on Drugs violated human rights norms. At times, such claims lacked specificity:

> The war on drugs is a human-rights crime. The war on drugs always was a human-rights crime. The war on drugs will continue to be a human-rights crime until such a time as the regime desists therefrom.
> ("Open Forum: Letters to the Editor,"
> *The Denver Post*, June 26, 2011)

Despite the public awareness raised by popular books like the legal scholar Michelle Alexander's (2012) *The New Jim Crow* and journalist Radley Balko's (2013) *The Rise of the Warrior Cop* that detail this phenomenon, claims about police militarization made up the least common theme in this frame with just 5%. Summarizing an interview with a protester, an author stated the following:

> She said drug policies are also run by a more "formal militarization" with an "increased use of SWAT units to knock down doors and break into alleged drug houses."
> ("Marchers Denounce Drug War in Mexico,"
> *The Blade*, September 6, 2012)

Political corruption and greed related to the prison industrial complex (i.e., a set of industries formed around the criminal justice system) comprised another theme that made up about 13% of total claims within the freedom and justice frame. These newspaper manuscripts argued that the War on Drugs is continued only because of the avarice of political elites and entrenched interest groups. The following quotations serve as some of the best examples of this argument:

> Hardly a week goes by without me seeing another think piece on the question: "Are we winning the war on drugs?" That depends on who "we" is. The war on drugs has certainly served powerful interests in our society. Between the drug war and the War on Terror, we've militarized police culture with SWAT teams, turned the Fourth through Sixth Amendments into toilet paper, and created the biggest prison-industrial complex in the world. From the standpoint of those who push the drug war the hardest, these are all – as Martha Stewart would say – good things.
> ("Two Cents: The 'War on Drugs' is really a war on you," *Deming Headlight*, November 21, 2011)

> There are simply many people whose livelihood depends on not winning this War-On-Drugs. Millions on both sides of the law depend on its continuance.
> ("Contra Costa Times Sunday Forum: Does the war on drugs need to change?," *Contra Costa Times*, August 6, 2011).

Mass incarceration and prison overcrowding constituted another recurring theme within this frame (28%), often without reference to the racial inequalities embedded within these concerns. Arguments focusing on this aspect of the War on Drugs emphasized the scope and growth over time of the incarcerated population, as demonstrated by the following excerpt:

> No other country in the world incarcerates as many people as the United States. There are almost 2.3 million people in America's prison system. The single greatest cause of the prison population growth has been the war on drugs, with the number of people incarcerated for non-violent drug offenses increasing more than twelvefold since 1980. This is simply unsustainable.
> ("Wyo. Pot Laws Harm Many," *Wyoming Tribune Eagle*, February 18, 2014)

Alongside statistics, claims in this theme used appeals to history or outlining how the current situation of mass incarceration came about:

> Beginning in the mid-1970s, legislators implemented "tough on crime" policies that created longer sentences, mandatory minimums, and new prison sentences for drug violations. Then the U.S. launched a "war on drugs" in 1982, at a time of declining drug use.
> ("Mass Incarceration hasn't made us safer,"
> *Deseret Morning News*, April 22, 2012)

Much like aspects of the fiscal frame, as the previous quote suggests, there is a clear role in statistical claims in a theme based on an issue of scale to articulate that a problem has become outsized. Global comparisons of the US prison population were also common. A news report quoting former US Attorney General Eric Holder stated, "the U.S. has 5 percent of the world's population but incarcerates almost a quarter of the world's prisoners, Holder said" ("Eric Holder proposes major shift in 'war on drugs,'" *The Christian Science Monitor*, August 12, 2013). Statistics were further employed to demonstrate the issue of overcrowding in prisons bloated by unnecessary arrests and long sentences for non-violent criminals:

> West Virginia has 1,700 more prisoners than prison beds. The Charleston Daily Mail reported in April that 21.9 percent of those incarcerated in West Virginia are in custody for drug related offenses, up 6.2 percent from 2004 with exponential growth predicted through 2020.
> ("War on Drugs has broken everything but the trade,"
> *Charleston Gazette*, July 10, 2011)

Another prevalent theme within the Freedom and Justice Frame saw the policies and practices of the War on Drugs in terms of violations of civil liberties. This theme made up about a quarter (24.5%) of all claims within this frame and largely related to the libertarian critiques of drug laws, perhaps best exemplified by journalist Jacob Sullum (2004), or even the classical arguments for personal liberty espoused by philosopher John Stuart Mill (1859). As these passages illustrate, central to these arguments are notions of freedom of choice, the value of the individual in society, and appeals to common sense:

> Socially and economically, the war on drugs is a disaster, but the real price tag has to be measured in terms of our liberties: drug use is a personal health issue, not a political one, and is therefore outside the scope of law — at least in a free country. A "free country" is defined as one in which the citizens have the right to act in any non-aggressive manner they choose.
> ("War on drugs: War on liberty, common sense," *Deming Headlight*, June 2, 2010)

> The United States should begin by allowing people to make their own choices that do not directly affect others. Of course, one's choice may indirectly affect others, but we should lean toward freedom of choice.
> ("Contra Costa Times Sunday Forum: Does the war on drugs need to change?," *Contra Costa Times*, August 6, 2011)

> Laws that are intended to protect us from our own bad choices are seen by many as not only an attack on liberty, but as an insult to their intelligence.
> ("Editorial: Letters to the Editor continued," *Richmond Times-Dispatch*, April 9, 2009)

Moreover, arguments within the Freedom and Justice frame more broadly, especially those based in minimizing the human costs of the "War on Drugs," often emphasized personal experience or common sense to present the claim maker as truthful:

> I am a Christian, a retired prison chaplain and the mother of a son who has had a long battle with drugs. I can speak from very close and personal experience.
> ("Letters to The Editor," *Austin American Statesman*, December 16, 2011)

The Functionalist Frame

By a large margin and across types of newspapers and manuscripts (i.e., op-eds, letters to the editor, news reporting), the **functionalist frame** was the most common, making up about 52% of the total claims critical of the War on Drugs. The term "functionalist" comes from the ideas of early sociologists such as Emile Durkheim ([1893]1984) and Talcott Parsons (1961). Functionalists argued that institutions such as the economy, education, religion, or the state provide important functions for society. They fulfill human needs and have express purposes and that they should facilitate order and stability. Functionalists deem social institutions dysfunctional if they fail to contribute to a more harmonious and secure society. The functionalist frame is a simplified version of this idea, pointing out that the War on Drugs has failed in its manifest functions. Manifest functions are the things that we expect institutions, policies, or organizations to do or what they explicitly claim to be doing (Merton 1949). In the case of the War on Drugs, its manifest functions include enhancing law and order in society and decreasing drug problems.

The seeming irony of the failure of drug policy to achieve any of its stated goals in a meaningful sense has often been fodder for comedy. A common punch line about the War on Drugs appeared in a 1998 headline in the infamous satirical newspaper *The Onion*. It simply stated, "Drugs win Drug War" (*The Onion* 1998).

Similarly, one of the most popular themes in this frame (26%) involved simply declaring the War on Drugs as failed, unwinnable, or lost. Often, claims within this theme relied on historical references to make use of the metaphor of war and its potential for victory or defeat:

> To say that the War on Drugs has been a miserable failure is a gross understatement. If World War II had been as unsuccessful as the War on Drugs we would all be speaking German right now and "heiling" somebody.
> ("Holder is Correct in Stopping Low-level Drug Cases,"
> *The Free Lance-Star*, August 16, 2013)

> In school they used to say the United States never lost a war. There was the Revolutionary War, then the War of 1812, then that thing with Mexico, then the Civil War (which, in the North, was considered a victory), the Spanish-American one, World War I and then, of course, World War II. Korea was something of a draw, and while Vietnam was a loss, there's no doubt that we could have won had we really wanted to. That leaves one war we have been fighting with everything we have

> since, it seems, time immemorial – and that we have lost. I am referring to the War on Drugs. Maybe it's time to throw in the towel.
> ("Just Say Yes?," *The Washington Post*, April 5, 1995)

Critics of the War on Drugs in mass media often claimed that its ineffectiveness was so apparent that its continuance was beyond all reason. Appeals to commonsense, at times to hyperbolic degrees, were common among claims within this theme, implying that the drug war is an obvious failure and thus its continuation defies basic logic:

> Common sense should have told us many years ago that prohibition would not work any better with drugs than it did with alcohol.
> ("Two Cents: War on Drugs Belongs to Us," *Deming Headlight*, July 25, 2011)

> This announcement is about as surprising as "Sun rises in east," "Dogs tip over trash cans," or "Mayor-elect denies allegations." If you've paid any attention to this topic, you don't need the former president of Switzerland, or more recently some retired police chiefs, to tell you that in this war on drugs, drugs have won.
> ("Time to Surrender 'War on Drugs' has been a miserable failure," *The Denver Post*, June 16, 2011)

Ideas of success and failure often hinge upon how well a policy or program performs in reducing or increasing certain behavioral trends. So, statistics on rates of drug use or drug-related social problems were employed to demonstrate that the War on Drugs lacks effectiveness:

> Drug warriors often point to the 1980's as a time when the drug war really worked because the number of illicit drug users reportedly fell more than 50 percent in the decade. But consider that in 1980 no one had ever heard of the cheap, smokable form of cocaine called crack or of drug-related HIV infection. By the 1990's, both had reached epidemic proportions in American cities. Is this success?
> ("An Unwinnable War on Drugs," *The New York Times*, April 26, 2001)

It is common to hear the War on Drugs described as a failure or even as the "failed drug war." Commentators often note quite legitimately that because drug problems and related issues remain and persist, it has not achieved its stated goals. Interestingly, newspaper content contained the word "failed" in overall descriptions of the War on Drugs as dysfunctional, almost as a shorthand for this idea:

> Decades of adherence to failed War on Drugs policies has helped make the U.S. the world's largest jailer, with only 5 percent of the planet's population but 25 percent of its inmates. Of the nearly 217,000 federal inmates, half are incarcerated for drug crimes, according to the Bureau of Prisons. Yet drug usage has risen 2,800 percent since the War on Drugs began in 1971.
> ("Editorial: Clemency is a good first step to ending failed War on Drugs policies," *Carlsbad Current-Argus*, May 10, 2014)

Not only did newspapers describe the War on Drugs as dysfunctional, but they also argued for reforms that would make drug policy more functional in achieving its goals of reducing drug use and crime. Less common themes included posing more functional alternatives to achieving the goals of reducing drug use in society, such as education (6%):

> Schools and community groups need to shift the focus from "scare tactics" to scientific facts about drug use. They need to teach students about harm reduction. If a teenager is vomiting, this is what you do. Don't be afraid to call 911. Don't drive drunk.
> ("Their Goal: Alternative to War on Drugs," *Providence Journal*, December 9, 2012)

> But a redirection of the millions spent to fight their import and distribution could be better spent on education, especially at an early age when such lessons leave a lasting impression.
> ("Editorial: Our View – Time for cease fire in the War on Drugs," *Moscow-Pullman Daily News*, April 24, 2012)

Many articles pointed out that drug use was a public health issue more than a crime issue because its natural consequences were primarily on the health and well-being of users. Therefore, another functional alternative routinely posed was public health-oriented approaches such as harm reduction and addiction treatment (18%):

> The London School of Economics report [...] outlines what its authors see as the "enormous negative outcomes and collateral damage" that have followed the militarized effort by governments who declared "war" on the illicit drug trade more than a generation ago and calls of those same governments to redirect taxpayer "resources away from an enforcement-led and prohibition-focused strategy" and instead focus on "proven public health policies of harm reduction and treatment" strategies for drug users."
> ("Washington: Nobel Economists Back Call to End Failed 'War on Drugs'," *Plus Media Solutions U.S. Official News*, May 7, 2014)

> The United States should treat drug addiction as a public health problem, not a law-and-order problem. Spending money on prevention and treatment works. Chasing down "drug kingpins" doesn't.
> ("A Bright and Shining Lie," *St. Louis Post-Dispatch*, February 10, 2001)

And, pointing out that an unregulated "black" market renders drug usage more damaging to both users themselves and society, a third functional alternative commonly proposed was legalization and regulation of the sale and distribution of drugs (13%):

> As long as this commerce is illegal, it is totally unregulated. Since we know that addicts will continue to buy drugs on the street, we also know that some will die from drugs that are either too potent or adulterated with other substances that could make them lethal. Is this really the intent of our drug policy? To invite users to kill themselves?
> ("We are losing the war on drugs," *Las Cruces Sun-News*, February 5, 2014)

> If you legalize drugs you instantly eradicate all of the crime associated with drug prohibition, with the illegal manufacture and distribution of drugs. You put the modern-day Al Capones

out of business. Then you can regulate and tax drugs, and educate people about them just like we do with alcohol and tobacco. An addict with a $ 1,000-a-week illegal habit who commits crimes to support it can fry his brain on $ 20-a-week of legalized drugs. His brain will be fried either way but he won't have to steal as much to do it. And he won't line the pockets of drug traffickers in the process.

("America's 'war on drugs' is not worth the mounting casualty list," *The Denver Post*, March 8, 1996)

The most common theme, however, at almost 35% of the total claims within this frame, went beyond merely critiquing the War on Drugs as failed or less functional than other approaches. Writers argued that it fails to generate law and order and, ironically, escalates the problems of crime and drugs. A major form of dysfunction attributed to the War on Drugs in newsprint was rooted in the observation that making an action a crime enables criminal enterprises. For instance, arguments often claimed that supply and demand in the illicit drug market are resistant to the policies and practices of the War on Drugs:

> Market forces and the law of supply and demand react quickly to indicators and likely future supply. Guzman's arrest should be seen for what it is: a triumph for law enforcement. But it will have no impact on the street prices for drugs. One more arrest of a kingpin or drug addict will not signal victory in sight.
> ("War on Drugs: Victory not in Sight," *Topeka Capital-Journal*, April 14, 2014)

> Drugs are not going away, and putting more people in prison is not the answer. But if we continue to wage this War on Drugs, let's add alcohol to the list and level the playing field. Buy a beer and spend a year in jail. Do you think that would stop Americans from drinking beer? If you do, you're living in a dream world. The same goes for the War on Drugs.
> ("Holder is Correct in Stopping Low-level Drug Cases," *The Free Lance-Star*, August 16, 2013)

From a functionalist perspective, the claims made within this theme are reasonable arguments against the assumed effectiveness of the War on Drugs as a drug control strategy. However, numerous claims within this theme relied on "group-based frames" or "racialized code words" (Hurwitz and Peffley 2005). These include Mexican cartels, inner city, street, or urban gangs, or Islamic terrorists evoking racialized images of danger, threat, and violence:

> This so-called "war" is funding the cartels and gangs, and their greatest nightmare is the legalization of drugs in the United States.
> ("Ending the War on Drugs,"
> *Deseret Morning News*, January 8, 2012)

> The War on Drugs [...] enriches criminals and terrorists. And it messes up our foreign policy. Meanwhile, drugs grow ever cheaper and more potent.
> ("Will he cross reefer Rubicon?,"
> *Providence Journal*, February 9, 2009)

> We have created some of the most dangerous criminals in the world by making the drug trade the gold rush of the past two generations. Street gangs all over the United States kill because of the drug wars. Many of the major cities have hospital emergency rooms that are overwhelmed with victims of gang warfare and drug overdoses.
> ("Letters to the editor,"
> *Carlsbad Current-Argus*, February 18, 2009)

> [The War on Drugs has] created and enriched violent gangs from Colombia to our inner cities. [...] Drug smuggling is an assault on America, and get-tough indignation is a natural response. We are not arguing for giving free passes to foreign smugglers. However, we need a new strategy that will reduce the demand for drugs and improve the safety and wellbeing of vulnerable Americans. What most of us want is less crime and less profit for foreign drug lords and urban gangs, and help for families, friends and neighborhoods afflicted by drug abuse and addiction.
> ("The Wrong War,"
> *San Jose Mercury News*, October 25, 1996)

These claims present a one-dimensional look at the issue of crime that reinforces the ideas of criminality as a racial trait that we saw in historical debates on race, immigration, and crime. They imply a threatening racial "other," a racial out-group associated with violence and crime, as the War on Drugs' most pressing danger.

Whites commit most drug crimes in the United States, likely to a disproportionate degree (Case and Katz 1991; Fairlie 2002; Fellner 2009; Ingram 2014). Yet, writers often emphasized opportunities for dysfunctional or dangerous people of color to engage in violent and criminal acts as the primary problem caused by drug prohibition. In linking the dysfunction of the War on Drugs with the assumed dysfunction of marginalized racial groups, these claims normalize the targeting of individuals and communities of color as drug criminals. This logic, while likely familiar to those of us who have engaged in discussions about drug policy, ironically mirrors many of the racialized moral panics used to develop and implement the War on Drugs in the first place.

The Racial Unfairness Frame

The least common frame in the newspaper dataset throughout each decade – manuscript type and newspaper type – was the **racial unfairness frame**, ultimately only contributing less than 9% of the total claims criticizing the War on Drugs. This frame criticizes the War on Drugs by focusing on its relationship with problems of racial inequality and oppression.

As shown in Chapters 2 and 3, research findings continually demonstrate several features of US drug policies and enforcement practices. First, they are mechanisms of racialized social control. Second, they contribute to systemic racism. Third, they take place within a racialized social system. And finally, they reproduce racial inequality. In contrast, the relatively minimal discussion of racial oppression within the public debate raises some interesting questions.

The most common theme within the racial unfairness frame involved claims that the policies and practices of the War on Drugs are racially biased (61%). Such claims often relied on statistics to demonstrate differential criminal justice outcomes for people of different racial background:

> He noted that according to one report, black male offenders receive sentences nearly 20 percent longer than those imposed on white males convicted of similar crimes.
> (Eric Holder proposes major shift in 'war on drugs',
> *The Christian Science Monitor*, August 12, 2013)

As it happens, its unintended victims have been disproportionately black. The stunning rise in incarceration rates for black men began after the nation became serious about stamping out recreational drug use.
("Let's Admit Drug War Has Failed," *Palm Beach Post*, December 31, 2007)

Largely because of the huge disparity in imprisonment for drug offenses, blacks are sent to prison at 8.2 times the rate of whites. Overall, one in 20 black men over the age of 18 is in a state or federal prison, compared to one in 180 white men.
("Report: War on Drugs Sends Blacks to Prison at 13 Times Rate of Whites," *The Washington Post*, June 8, 2000)

In newspapers, commentators routinely noted that "the statistics are alarming" ("Mass incarceration hasn't made us safer," *Deseret Morning News*, April 22, 2012) or "the disparities are particularly striking" ("Report: War on Drugs Sends Blacks To Prison at 13 Times Rate of Whites," *The Washington Post*, June 8, 2000) before launching into an inventory of percentages and ratios. Claims within this theme were also often listed off briefly rather than fully explained or extrapolated. Comments included these references to statistics as merely one further reason why the War on Drugs is problematic among a catalog of largely nonracial critiques. One article included toward the end after discussion of many other issues: "Moreover, there is an unacceptable racial bias in the enforcement of marijuana laws" ("Wyo. Pot Laws Harm Many," *Wyoming Tribune-Eagle*, February 18, 2014).

Another theme within the racial unfairness frame involved claims that black communities (or communities of color more broadly) happen to be uniquely susceptible to the problems of drugs, policing, and crime connected to the War on Drugs. These claims made up almost 14% of the total claims in this theme. In particular, this theme centered on a claim about unfairness in outcomes and opportunities, similar to what criminologists call strain theory, by explaining crime as a product of a lack of legal opportunities for status and resources (see Merton 1932). Yet they often relied on the trope of drug criminals as predominantly black or Latino:

> The attractive illegal livelihood relieves men of the need to develop skills that would provide stable legal incomes. To those

who argue that there's a shortage of jobs for black men, he says that is refuted by the black immigrants who thrive in America. It is often said that because immigrants have a unique initiative or 'pluck' in relocating to the United States in the first place, it is unfair to compare black Americans to them. However, the War on Drugs has made it impossible to see whether black Americans would exhibit such 'pluck' themselves if drug selling were not a tempting alternative.
("Save Black America by Ending the Drug War," *The Union Leader*, March 18, 2011)

Before about 10 years ago, cocaine was a drug rarely used or dealt with by poor blacks. The cost of it limited its use to those of more means. But once it became accessible, it became a scourge on many poor black neighborhoods across the country, leaving behind a trail of addiction and carnage wherever it went.
("Laying Blame for Epidemic of Crack Use," *St. Louis Dispatch*, September 24, 1996)

However, not all claims in this vein connected drug crime with black or Latinx communities. Other examples of this theme acknowledged that people of color do not contribute to most drug crimes. They instead attempted at accounting for how the nature and qualities of drug crimes committed in low-income urban communities render black citizens more likely to face arrest and punishment:

> Minorities especially suffer from the fact that inner-city drug dealers tend to congregate on the street, not indoors as in more affluent suburbs where the bulk of illegal substances is actually consumed. Because street-level dealing is so easy to spot (as television discovered long ago), police target it with "buy-bust" arrests that pump up their statistics and appease the public's demand for action. [...] Similarly, no resident of Los Angeles doubts for a moment that police are more likely to make drug busts in Watts than in the Hollywood hills.
> ("How Our War on Drugs Shattered the Cities," *The Washington Post*, May 17, 1992)

The above examples highlight important aspects of the racial implications of the War on Drugs. Yet, they also leave much to be desired from a sociological perspective. However, claims about the War on Drugs and its relationship with racial inequality that employed more systemic or structural understandings appeared much less often. One relevant theme in this vein was that the War on Drugs is motivated by racial attitudes, including animus and apathy toward people of color. This theme only accounts for 8% of the total claims within the racial unfairness frame and only five times total within the dataset. Opinion piece authors used this theme to argue that support for the War on Drugs was not based on providing sensible solutions to drug problems, but rather was intrinsically connected to racism. In news reporting, this theme was not remarked upon by journalists themselves. It only appeared as quotes or summaries of comments from prison or drug reform activists or people negatively impacted by the War on Drugs. This theme usually appeared in reference to the historical roots of the War on Drugs and the construction and strategic use of racialized moral panics[3]:

> President Richard Nixon was the first to use the phrase "war on drugs" in 1969. After Watergate, it was revealed his motives were purely political. Campaign leaders also coined the term "silent majority" to appeal to the millions of white, middle-class suburban voters who were horrified by student protests against the war in Vietnam. The war on drugs policy resulted in federal law that made substances like marijuana illegal (typically, drug laws are made at the state level). This appealed to those "silent majority" voters who were angry about what they saw as drug orgies at events like Woodstock and who feared militant urban blacks with whom they associated drug use.
> ("White Flag on Drugs," *Bangor Daily News*, May 17, 2010)

Another critique of the War on Drugs from a perspective of how it reproduces racial stratification entails pointing out that it is a cause of racial inequality in society. This theme was slightly more common at almost 17% of the total claims in this frame, predominantly focusing on the adverse effects of these policies and practices on black and Latinx families and communities:

> The nation's drug policies have devastated the minority community by incarcerating thousands of youth for minor

possession charges, torn families apart and turned law enforcement against the very communities it was trained to protect.
("Their Goal: Alternative to War on Drugs," *Providence Journal*, December 9, 2012)

As you must know, the War on Drugs has been, in effect, a war on black men. Though whites are the nation's biggest users and dealers of illicit drugs, blacks are the ones most likely to be jailed for drug crimes and to suffer the disruption of families and communities that comes with it.
("Please end America's most wasteful war," *The Herald-Sun*, June 18, 2011)

While there was some discussion of these issues, the intense focus on racial bias, or different treatment or outcomes for different groups, within this frame only communicates part of the story discovered by social scientists, historians, and legal scholars. The War on Drugs is not simply biased but part of the social system that reproduces racial inequality in US society. In other words, it is part of a large-scale process that helps maintain disparities in symbolic and material resources between racial groups. However, only 2% of claims discussed the War on Drugs as a product of racial attitudes or a cause of racial inequality. Without direct knowledge, intentional research, or otherwise coming across such information, someone could be deeply involved in this debate and yet never actually engage with some of the most important aspects of this issue: its historical genesis in racialized moral panics, its role in the maintenance of racial oppression, and its impacts on individuals, communities, and families which contribute to not only undue hardship but overall outcomes of racial inequality.

Conclusions and Unanswered Questions

Identifying the frames that people used to argue about the War on Drugs in newspapers is only one step toward understanding the big questions about society that motivate this book. Understanding how often frames appeared and what they emphasized helps us understand the effects of agenda-setting within the media, including the definition of the War on Drugs as a social problem. So, looking at how newspaper content framed criticisms against the War on Drugs over the past three decades, what did we learn?

The functionalist frame was the most widely employed way of critiquing the War on Drugs within the debate. This frame presented the War on Drugs as a failed approach that didn't achieve its goals. Other claims suggested that it may have amplified other social problems such as drug abuse and violence. Often, these arguments presented the dysfunction of the War on Drugs by suggesting that it exacerbated illegal activity. Through suggestive or covert language, many of these claims implied that the criminals and immoral characters abetted by the policies and practices of the War on Drugs were people of color. Almost none of these claims mentioned race outright, but, as Hurwitz and Peffley (2005) found out in their study on "group-centered frames," such subtle insinuations are nonetheless powerful. The way that ideas, activities, and policies are associated with different groups (even in indirect ways) impacts how people make claims about themselves, others, and society while debating contested social issues.

Arguments about the fiscal soundness of the War on Drugs or its general implications for a free and just society were pervasive. Criticisms that the War on Drugs wastes taxpayer money, negatively affects the economy, violates individual liberty, or overfills our prisons were, on their face, colorblind. The previous chapters of this book highlighted the relationship between racial oppression and drug laws. So, what about the racial implications of the policies and practices of the War on Drugs? Why did so many arguments mention racial bias without discussing why or how this racial bias takes place? And why were these other frames so common? These are questions that just looking at this dataset alone can't answer. Fortunately, we have some powerful theories and even more data to help us tease out these lingering "why" questions. In the next chapter, we'll examine these results by introducing the concepts of racial silence, resonance, and code words, and by looking at how they might influence debates in society over racialized social issues.

Discussion Questions

1 Which arguments discussed in this chapter were familiar to you? Where did you learn about them? Do you agree with them? Explain?
2 Why is it so challenging, particularly for whites, to talk about racism? Explain.

3 Think of a contested social issue commonly discussed in the news today. How many different frames can you identify for how this issue could be discussed? What are the benefits and limitations of each of these "frames?"

Notes

1 Per Pew Research Center survey data from 2014, 42% of Republicans supported punitive approaches to drug use as opposed to just 18% of Democrats.
2 See Rosino and Hughey (2016) for a more detailed explanation of the research methods used for this analysis.
3 See Ian Haney-Lopez (2014) on "strategic racism."

Chapter 4

Debate Dynamics
Racial Silence, Resonance, and Code Words

Debates in the media include narratives defending the way things are and those criticizing its problems and downsides (Schudson 2011). In the previous chapter, I explained how agenda-setting and framing take place in the media. We also saw how arguments against the War on Drugs in newspapers framed this issue. While we now know the "what" in terms of media framing, we still need to understand the "why." Why were uses of the "functionalist frame," depicting the War on Drugs as a failure or cause of crime and drug abuse, so prevalent? Why were applications of the "racial unfairness frame," especially those that related to sociological understandings of racial inequality, so rare? In this chapter, I will introduce several concepts and theories that explain this distribution of frames in the newspaper debates on the War on Drugs.

An essential aspect of the debate that can help us understand these trends is that the War on Drugs is a racialized social issue as well as a contested one. **Racialization** is "the extension of racial meaning to a previously racially unclassified relationship, social practice, or group" (Omi and Winant 2014:110). The War on Drugs is racialized because it is acutely associated with racial meanings. These include moral panics about people of color and the racial ideologies that have animated the growth and defense of punitive drug policies. Preconceptions about racial groups continue to influence attitudes and public policies that distinguish between drug users of different racial, ethnic, and class backgrounds. We cannot understand the debate about the War on Drugs without considering the following: (1) how Americans learn to think and feel about various racial groups; (2) how they define racism; and

(3) how they recognize and account for ongoing racism and racial inequality in the United States. Extending our understanding of contested social issues to racialized contested social issues requires an expansion of our conceptual toolkit.

Racial Silence

In my analysis of the debates in the newspapers, I found little variation in the distribution and features of frames across newspaper types and decades. One of the most consistent trends was that claims critical of the War on Drugs on the grounds of racial unfairness constituted the least common frame across all the decades and manuscript and newspaper types in the dataset. This lack of direct discussion about race and racial justice within the overall debate in newspapers suggests the presence of **racial silence**.[1] Given that the War on Drugs is a highly racialized social issue, racial silence is somewhat counterintuitive. However, it is not unprecedented.

Ruth Frankenberg (1993), a whiteness and women's studies scholar, interviewed white women about racial issues and their daily lives. She found that they routinely engaged in practices she called color evasiveness or power evasiveness. Color evasiveness is an unwillingness to acknowledge the role of racial categories in shaping people's everyday lives. Power evasiveness is an unwillingness to recognize the differences in social, economic, and political power held by different racial groups. Subsequent research has further demonstrated that white people often engage in avoidance (Doane 2003; Lewis 2003b) or evasion (Steinberg 2007; Mueller 2017) in discussions of racial issues to sidestep the contentious and controversial seeming nature of racial inequality and its social causes.[2]

However, racial silence does not just result from intentional strategies that people use for making their claims seems less controversial. It also reflects their knowledge of racialized events and issues like the War on Drugs. France Winddance Twine (2010), a sociologist and whiteness studies scholar, developed the concept of **racial literacy** to explain how white British mothers prepared their children of multiracial heritage for racism by explicitly discussing race and racism with them. These white mothers provided them with a vocabulary, set of concepts, and historical knowledge. They countered evasion and silence by engaging in training that allowed their children to perceive, understand, and discuss the role of racial categories and racial oppression in society and their everyday lives. But not everyone possesses racial literacy.

When trying to understand racialized events and issues, people often ignore the historical and current social arrangements which advantage whites and disadvantage people of color. Take the racial wealth gap, for instance, where whites have an enormous lead over people of color due to past and ongoing policies, government programs, and access to social networks and opportunities (Oilver and Shapiro 1995). Jennifer C. Mueller (2017), a sociologist, examined how US college students responded to an assignment in which they talked to relatives about how their families' current social standing reflected access to these types of wealth-building opportunities.[3] Mueller (2017) found that even in the face of clear evidence, many white students avoided full recognition that they benefited from these arrangements due to an epistemology of ignorance. **Epistemology** refers to how people produce knowledge.

Unequal societies, like the United States, organize people into social hierarchies around categories like race, class, and gender. The **social position** of a group in that hierarchy reflects relationships with other groups and associations to institutions that provide access to resources (Collins 1993). Accordingly, social group members develop a "sense of group position" (Blumer 1958; Hughey 2011). Charles Mills (2007), a political philosopher who conceptualized the term **epistemology of ignorance**, argued that whites often have gaps in their knowledge of the experiences of people of color and the existence of racial oppression due to their dominant social position. And this lack of knowledge allows them to normalize this position. Not thoroughly learning how wealth connects to public policies that distributed real estate, voting rights, education, and other resources along racial and class lines is central to the maintenance of an epistemology of ignorance. And as Stephen Steinberg, a sociologist, wrote, "an epistemology of ignorance speaks through silence" (2007:43).

For some readers, especially those who are white, it may cause discomfort or even defensiveness to think about how racial oppression impacts the lives and minds of white people. But as Amanda E. Lewis (2004:626), a sociologist, argues, "in a racialized social system all actors are racialized, including whites." What does it mean to say that whites are racialized? Racial categories such as "white" or "black" don't reflect self-evident or primordial realities, but rather social inventions (Graves 2001; Obasogie 2013).

European elites invented the racial category of "white" in the US colonies in the 17th century (Allen 1997). In subsequent centuries, political and economic elites, along with those who became classified as "white," embedded this new distinction into laws, institutions, and the informal rules of people's everyday routines (Allen 1997; Gossett [1965] 1997; Coates 2003b; Rosino 2017).

The invention of "whiteness" created new forms of social division that rationalized the social oppression and economic exploitation of groups placed outside of this category (Allen 1997; Roediger 1991). Over time, even people in the lower socioeconomic classes who became classified as "white" received advantages in public life and often developed a sense of superiority (Du Bois 1935; Roediger 1991). This arrangement created incentives for many of those endowed with "whiteness" to maintain racial divisions and social hierarchies (Du Bois 1935; Roediger 1991). And because this category has tended to correspond to arbitrary social and economic advantages, which ethnic groups count as "white" has been a source of conflict for much of US history. For instance, in the 1920s, two major US Supreme Court cases weighed whether people from "non-white" ethnic groups counted as "legally white" and therefore qualified for citizenship and other benefits and advantages (Haney López 2006).

Despite this historical and contemporary context, people tend to imagine racial categories and racial oppression as only impacting the lives of people of color. As Ruth Frankenberg (1993:6) pointed out, racism is not just "something that people of color have to deal with in a way that bears no relationship or relevance to the lives of white people." To make these relationships visible and, therefore, better understand how meanings of race and practices of racism impact everyone, researchers in **whiteness studies** analyze how people perceive and live out that racial category in their everyday lives (Twine and Gallagher 2008). The reality of the invention and social advantages of whiteness contradicts the seeming invisibility of its effects to many whites. Digging deeper into the causes and consequences of this paradox can help us comprehend the racial silence in the print media debate over the War on Drugs.

By bringing to light the racial silence in the War on Drugs debate throughout a wide range of time periods, regions of the United States, and newspaper formats, the findings presented in the last chapter demonstrate the cumulative effects of these practices and processes over the contours and norms of debates. Going back to our discussion in Chapter 2 on media and power, we might say that the dominant framing of the problem of the War on Drugs does not include consideration of its implications for racial equality. For instance, a letter to the editor writer critiqued an op-ed quoted in the previous chapter that argued the War on Drugs targets black men by stating:

> Pitts was fine up to the point where he decided to pull out the race card. I have to disagree with him when he says that "the war on drugs has been, in effect, a war on black men." Please, let's stop having one group of people or another blaming others for misfortunes they have largely brought upon themselves. It is important for the young people of all races to stay in school, get a job and resist the gang life. In my view, people should learn to live their own lives without blaming everyone else for their circumstance.
> ("Letters to the Editor," *Herald News*, 2011)

What does it mean to say that someone pulled out the "race card"? Americans often use this phrase to dismiss claims of racism or discrimination as unreasonable or outside the bounds of discussion. Some types of claims are more likely to be viewed as improper usage of the "race card" than others. Arguments about the War on Drugs that involved thinking about racial inequality as tied to the social system were exceptionally rare in the debate in newspapers. Ashley W. Doane (2006:262), a sociologist, argues that there are "clear lines of conflict between the 'color-blind' view of racism as prejudice or hate and the alternative perspective of racism as a more structural phenomenon embedded in American society." For instance, mass media messages often critique the War on Drugs in ways that mention the existence of racial bias. But they often fail to connect this bias with the structure of society or identify the systems and processes responsible for producing racial inequality. Because of this omission, they leave open the interpretation that racial bias in the War on Drugs is a product of a few individually biased people or a naturally occurring phenomenon. And even more claims simply pointed out numerical racial disparities in the criminal justice system without noting their causes or effects.

Institutional racism, how the rules and routine practices of social institutions produce racial disparities, can be seen in the War on Drugs via how institutions such as criminal justice and courts are organized (Provine 2007). Which groups become targets of police surveillance further exemplifies institutional racism (Lynch et al. 2013). Though institutional racism is often less explicit and it can be more challenging to identify specific individuals who are responsible, it is still "no less destructive of human life" (Ture and Hamilton 1967[1992]:1). In contrast to instances of individuals committing overt acts of racism, institutional racism "originates in the operation of established and respected forces in the society, and thus receives far less public condemnation" (Ture and Hamilton 1967[1992]:1). Scholars have described similar forms of discrimination and inequality which take place through the structure of society, the decisions made by people in positions of authority, and the rules and impacts of institutions as "internal colonialism" (Blauner 1972), a "racialized social system" (Bonilla-Silva 1997), or "systemic racism" (Feagin 2001).

As noted in the first two chapters, there is well-documented evidence of institutional racism in policing and the legal and prison systems that implement the War on Drugs. However, within newspapers, commentators avoided stating the racially unequal outcomes of the War on Drugs as a product of institutional racism in clear terms. Pointing out disparities is essential. But the audience of these newspapers were often left to come to their own conclusions regarding the causes and meaning of these disparities. The more critical and systemic a claim about racial unfairness was, the less likely it was to appear in newspapers. Even while pointing out racial unfairness or bias, very few arguments mentioned racial attitudes or motivations, the arrangement of institutions and rules, and the contributions of these to the overall outcome of racial inequality. Perhaps, such claims do more to puncture the veneer of racial silence in their explicit indictment of society and institutions.

Arguments in defense of the War on Drugs employed even more racial silence. In all the newspaper manuscripts in the sample, only one pro-War on Drugs claim mentioned race or racial categories. It used racial language to defend the targeting of specific neighborhoods as natural or inevitable in response to mounting and conclusive evidence of racial bias in the War on Drugs:

> "What is a police officer supposed to do if he sees a black guy on the street breaking the law, not arrest him?" asks James Pasco of the Fraternal Order of Police. "The people who live

in those areas . . . are the victims, and they are usually of the same ethnicity as the perpetrators."

("Study: War on Drugs is stacked against blacks," *USA TODAY*, 2000)

Similar practices, such as focusing on the nonracial implications of a racialized policy or event, can be found in historical debates (see Henricks 2017). But the emergence of racial silence, as we see it in the contemporary discussion on the War on Drugs, also reflects modern ideas, institutions, and ideologies. According to this contemporary logic, we already live in a free and equal society. Accordingly, even just noticing racialized issues and discussing them in overt terms is perceived by many whites as itself a form of racism worthy of stigma and derision (Doane 2006).

People often observe racially unequal outcomes, but disagree about the reasons they exist. Eduardo Bonilla-Silva (2014), a sociologist, has explained how the maintenance of racial inequality in the contemporary United States is often rationalized or denied through superficially "nonracist" frames. Without conforming to traditional definitions of racism, many whites in modern US society justify racial inequality by depicting it as inevitable and tied to cultural differences. They often argue that racial discrimination is no longer prevalent or claim that social and political changes to address racial inequality are themselves unfair or problematic. Given the dominance of this style of talking and thinking about racial inequality or **colorblind ideology** (Bonilla-Silva 2014), it is perhaps understandable that racial silence is typical within both supportive commentary and critiques of the War on Drugs.

Colorblind ideology, the contemporary dominant **racial ideology**, is composed of sets of ideas that people use to interpret racially unequal outcomes in society as just, natural, or reasonable (Bonilla-Silva 2014). The concept of **ideology** comes from Karl Marx and Freidrich Engels (1978[1845–1846]:172), 19th-century German historians and philosophers, who wrote, "the ideas of the ruling class are in every epoch the ruling ideas, i.e., the class which is the ruling material force of society, is at the same time its ruling intellectual force." This theory suggests that throughout history, the most influential people have produced the prevailing or dominant ideas of a given period. Therefore, these ideas rationalize or hide certain conflicts, problems, and inequalities. Because of their access to material resources, dominant groups tend to hold more influence over the ideas that come to be accessible and widespread in society. These widely held ideas, therefore, often represent the worldviews and interests of the dominant group.

Given its relationship to widespread and dominant ideas about racial inequality, racial silence on the issue of drug policy and enforcement extends beyond the debate in recent newspapers. Racial silence on the subject of drug policy extends to policy reform organizations such as Marijuana Policy Project and NORML that argue against the War on Drugs yet have historically omitted its role in reproducing racial inequality (Hart 2013b).[4] Moreover, television news reporting on drug policy often contains jokes and moral panics rather than analysis of its social and especially racial implications (Trujillo 2012). As these examples illustrate, Americans regularly encounter viewpoints on drug policy that focus on seemingly nonracial components or present the racial outcomes and implications of the War on Drugs as natural or legitimate.

Racial silence also extends beyond discussions of drug policy. Consider how often conversations about all kinds of contested social issues take place online in venues like social media. Racial silence is the norm in large parts of the digital sphere. Recent survey research by the Pew Research Center (2016a) found that most white Americans rarely, if ever, encounter and post material about racial issues on social media. Being surrounded by the racial silence present in many areas of our society prevents people from gaining knowledge. It enables whites to ignore information that contradicts their sense that their place in the social hierarchy and its impacts on their daily lives are natural and can be taken-for-granted.[5] And as noted by Eviatar Zerubavel, a sociologist (2006:9), "silence often involves an unspoken conversation." Zerubavel (2006:9) points out that:

> Being silent [...] involves more than just absence of action, since the things about which we are silent are in fact actively avoided. The careful absence of explicit race labels in current American liberal discourse, for example, is indeed the product of a deliberate effort to suppress our awareness of race. Ironically, such deliberate avoidance may actually produce the opposite result.

What are the unspoken conversations hidden by racial silence? For one thing, racial silence might obscure the racial justice implications of the War on Drugs from a large proportion of Americans. Considering widespread skepticism toward claims of racism and discrimination (van Dijk 1992; Doane 2006; Bonilla-Silva 2014) and highly racialized representations of crime in mass media (Robinson 2000), the omission of racial justice claims might also serve to appeal to the worldviews and perspectives of predominantly white audiences. The idea of resonance is thus an important one.

Resonance

Another way of understanding why specific claims were more or less frequent in arguments about the War on Drugs in newspapers is by thinking about whether or not they are likely to resonate with readers. **Resonance** is a crucial feature of how we perceive and consume cultural objects like newspapers. Understanding whether and how particular messages resonate with their audiences is key to answering the question of how framing and agenda-setting operates as strategies in newspapers and other forms of media.

What does it mean to say that something resonates with you? Think of a time when you heard a friend tell a story or describe an experience, and it elicited a strong feeling of understanding. You might have even found yourself nodding vigorously or replying with something like, "I know exactly what you mean." Have you ever thought about why? Would someone else's story, if they lived in a completely different social and cultural context or had an opposite worldview, have resonated with you so deeply? In a broad sense, our past experiences, perspective, socialization, and knowledge can impact how we feel about new ideas and narratives, what they mean to us, and whether we relate to them.

Mass media content is often intentionally crafted to resonate with specific audiences and their perceptions of US society. In one sense, this is a matter of necessity. Newspapers that print stories that their audiences can't relate to run the risk of going out of business. It also reflects the assumptions of those who produce mass media. Media organizations with a disproportionate majority of white employees receive less input reflecting the views and experiences of people of color (Downing and Husband 2005). Dominant groups find themselves in a position to create mass media content based on their ideologies. Even within stories and events ostensibly about people of color, mass media content routinely depicts the perspectives and experiences of whites as idealized, universal, and normative (Downing and Husband 2005; Hughey 2009, 2010).

Similarly, assumptions about what types of frames will resonate with intended audiences shape mass media content on racialized social issues such as the War on Drugs (Rosino and Hughey 2017). Media commentators, average citizens, and journalists often write opinion pieces or letters to the editor meant to persuade predominantly white audiences by appealing to their worldview or interests. But such assumptions don't just impact opinion-based content intended to be persuasive. Journalistic reporting on this topic, assumed to provide neutral or purely descriptive accounts, also avoided seemingly controversial or politicized language and concepts such as racism or racial oppression unless describing the views of others. Newspaper articles often go well out of their way to avoid using terms that they view as controversial and, instead, use awkward phrases such as "racially tinged" or "racially charged" (Glickman 2018). Yet, in doing so, they omit these relevant social realities. As Lawrence B. Glickman (2018), a historian, points out, such euphemisms "suggest that race is a fact—something that can be highlighted in a neutral way—rather than an ideology, a tool of oppression."

As Stuart Hall and his colleagues (1978:55) argued, "the media are among the institutions whose practices are most widely and consistently predicated upon the assumption of a 'national consensus.'" The assumption among media producers of a widespread agreement about the crucial aspects of an issue impacts the mass media content about that issue or event. The content of the debate over the War on Drugs in print media reflects "what the audience is assumed to think and know about the society" (Hall et al. 1978:57). And regardless of the accuracy of these assumptions, they produce self-fulfilling prophecies. Mass media content reflects certain expectations about what will resonate with audiences, but that content also impacts how people view contested social issues.

To better understand resonance, let's return to an example from our previous discussion of racial silence. Describing the War on Drugs as a "war on black men" did not correspond to or resonate with the experiences and worldview of many nonblacks, like the letter writer who responded. Perhaps framing critiques of the War on Drugs in terms of fiscal soundness or its effectiveness at reducing crime rates would have elicited a much friendlier response. However, the likelihood that media content holds resonance does not merely reflect a private relationship (between the individual and the material) or even a social relationship (between content and audience). Instead, media content such as newspaper articles shares a "public and cultural relation among object, tradition, and audience" (Schudson 1989:170). In other words, media content on the War on Drugs resonates when it aligns precisely with audience members' understandings of issues such as race, crime, politics, and drug laws. As, Michael Schudson (1989:169), a sociologist and media scholar argued:

> The relevance of a cultural object to its audience, its utility, if you will, is a property not only of the object's content or nature and the audience's interest in it but of the position of the object in the cultural tradition of the society the audience is a part of. That is, the uses to which an audience puts a cultural object are not necessarily personal or idiosyncratic; the needs or interests of an audience are socially and culturally constituted. What is "resonant" is not a matter of how "culture" connects to individual "interests" but a matter of how culture connects to interests that are themselves constituted in a cultural frame.

This idea of resonance is nuanced. So, let's break it down a bit with an example. To bring it back to the op-ed and letter to the editor, it was not merely that calling the War on Drugs a "war on black men" didn't line up with the personal interests of the respondent. They agreed that the War on Drugs should end. Instead, this claim failed to correspond or resonate with the interests considered ideal by influential members of the racial group to which the letter writer belongs. This group also includes those who control the cultural traditions of the media. These idealized interests include a few different dominant ideas. First, it implies personal responsibility (i.e., "people should learn to live their own lives without blaming everyone else for their circumstance"). Second, it demonstrates the belief that the problem of "racism" was solved during the Civil Rights era, and that racial discrimination is no longer a pervasive problem, and that, therefore, unequal outcomes are a product of individual behavior. These statements represent certain ideals. And these ideals correspond to the interests of white elites in ignoring institutional racism in the War on Drugs.

In other words, resonance often relates to **cultural ideals** or dominant ideas about the things that should happen. As evidenced by one of the earliest sociological studies, cultural ideals often align with group interests. Max Weber (1958 [1904–1905]), a late 19th- and early 20th-century German sociologist, showed how, during the industrial revolution, the "Protestant ethic," a set of concepts, stories, and religious beliefs that idealized hard work, success, and the accumulation of wealth, became the "spirit of capitalism." These ideas, principles, and narratives aligned with the interests of an emerging class of industrial capitalists. Weber (1958[1922–1923]:280) famously wrote: "Not ideas, but material and ideal interests, directly govern men's conduct."

The messages within cultural products like newspapers that carry the most extensive resonance with audiences also tend not to challenge the interests of white elites in maintaining their dominant group status. Arguments that ignore issues of racism, structural inequality, mass incarceration, and those which suggest that the "real" problems of the War on Drugs relate to the need to subdue threats, often represented as people of color, to the maintenance of order in society are the most common. Even though racial silence was the norm, ideas about racial groups still existed in the debates that took place within the pages of these newspapers. To understand the frequencies of frames in our newspaper sample, alongside the meaning or content of framed messages in the media and how they resonate with audiences, we should also pay attention to how people express them.

Code Words

How can people discuss a racialized social issue while still appearing to maintain racial silence? One of the most common themes in the sample, part of the functionalist frame, often suggested the relevance of racial groups. However, it routinely fell short of naming them directly. Claims that the War on Drugs is dysfunctional because it escalates the problems of crime and drugs commonly suggested racial imagery but did not explicitly state it. Instead, they used **code words** to stand in for the explicit communication of ideas and assumptions about racial groups, especially people of color.

Open expressions of overtly racist language and ideas and explicit defenses of racial inequality and white supremacy have long been fixtures of public life in the United States (Du Bois 1953; Jordan 1968; Gossett [1965] 1997; Feagin 2001; Anderson 2016).

However, in the wake of significant cultural, social, legal, and political transformations in the 1950s and 1960s, public disgrace and scandal are now more likely to accompany open and public displays of racism in contemporary US society. But this change does not mean those racial (or racist) ideas and positions have themselves necessarily diminished. Bonilla-Silva (2002:46), for instance, wrote that "because post-civil rights racial norms disallow the open expression of direct racial views and positions, whites have developed a concealed way of voicing them."

In the aftermath of the 2017 election of Donald J. Trump, examples abound suggesting that the public expression of hate speech, racial slurs, and epithets became more frequent and less stigmatized among some segments of the white population (Bredderman 2016; La Porte 2016; Okeowo 2016). In just the first ten weeks after the election, the Southern Poverty Law Center reported 867 incidents of harassment and intimidation based on race, religion, gender, or sexuality (2016). Before this election, the use of code words was more common. And yet, overall in comparison to previous periods, overtly racial slurs and epithets are still widely stigmatized and less commonly used in public (Bonilla-Silva 2002, 2014).

In recent times, whites wishing to express racist views in public but anonymous spaces such as social media have taken extreme measures of code word usage. Take the following example. An article in the *Miami Herald* (2016) quipped, "If you come across a tweet that seems angry about Skype, Yahoo and Google, it may be far more sinister than a complaint about online service." White supremacists created a code to communicate racial epithets on social media without censorship: "Google stands for the N word; Skype for an anti-Semitic slur that nearly rhymes; Bing for derogatory terms about Asians; Yahoo for Mexicans; Skittles for Muslims[6]; and so on" (*Miami Herald* 2016).

It is one thing to note a lack of overt racial and ethnic slurs in conversations in the contemporary public sphere, but even beyond the concealment of overt racial epithets, direct mentioning of racial groups, especially in a negative light, is often condemned in the present era. Ian Haney Lopez (2014), a legal scholar, argues that coded racial appeals or what he calls dog-whistle racism have been advantageous for politicians in the United States. Code words activate racial imagery or meanings in the minds of audiences while enabling the speaker to deny any racial insinuations or motivations.

Throughout the country's history, US political leaders have frequently made overtly racist statements, many of which were even considered extreme for their time. These statements often helped them gain the support of white citizens by appealing to their sense of group interest or fear of racial others. George Washington (1783), the first US President, wrote that "Indians and wolves are both beasts of prey, tho' they differ in shape." The third President, Thomas Jefferson, claimed that blacks were "pests in society," "as incapable as children of taking care of themselves," and "inferior to the whites in the endowments of body and mind" (Finkelman 2012). Two decades later, the seventh President, Andrew Jackson (1833), stated in a public address that indigenous people in the United States "have neither the intelligence, the industry, the moral habits, nor the desire of improvement which are essential to any favorable change in their condition." He even went on to describe whites as "a superior race" (Jackson 1833). Almost a century later, the twenty-third President, Thomas Woodrow Wilson, openly praised the Ku Klux Klan and their agenda to enforce racial segregation through violence (Matthews 2015).

However, code words and subtler expressions of racial prejudice began to play a more serious role in American politics during the Civil Rights era, as a consensus began to form that overt expressions of racial prejudice were old-fashioned, immoral, or inappropriate. In the 1960s, Alabama politician George Wallace discovered that he could use nonracial language to take stances on racial issues such as framing opposition to racial integration as a matter of states' rights rather than the intrusion of blacks into white spaces (Haney López 2014). Other politicians, such as Richard Nixon, then picked up on his strategy. Nixon used coded racial language to win the presidency by using "law and order" as a code for racial resentment against the protests and civil disobedience employed by the civil rights movement (Haney López 2014).

While the potential for a backslide into more overt racial rhetoric lingers, the use of coded racial language by political leaders from both major parties remains pervasive (Rosino and Hughey 2016). Take, for example, the issue of welfare. Welfare and public assistance programs more broadly have been dismantled in contemporary US society through association with images and narratives of black people as idle, reliant, and ultimately undeserving of public investment (Gilens 1999; Neubeck and Cazanave 2001; Collins 2009). Republican politicians repeatedly used code words during the election and tenure of President Barack Obama by associating his presidency with terms like "welfare," "entitlement," and "food stamp" (Hughey and Parks 2014). Democratic politicians have also used code words to their benefit, such as President Bill Clinton using racial code words such as "government dependency" and "welfare abuse" to push for welfare reforms in the 1990s that arguably left millions of Americans in poverty (Haney López 2014).

Code words resonate with many whites' racial views while conforming to contemporary racial norms; they also become part of the public discourse and allow everyday people to talk about racial issues while observing a vow of racial silence. Political scientist Tali Mendelberg (2001) demonstrated through a series of innovative social experiments that once the racial meanings of coded racial messages are made explicit or decoded, they lose their effect. Arguably this occurs because these messages then break the norm of racial silence or colorblindness. Claims that "food stamp recipients," "illegal aliens," or "inner-city criminals" are morally pathological have different effects on white audiences than outright claims about black or Latinx Americans.

Alternatively, as demonstrated by Hurwitz and Peffley's (2005) research on group-based frames, these racially coded claims might have different effects than racially neutral arguments about policy issues. Measuring this effect would be difficult, however, given that issues like social programs, immigration, and crime are often already racialized in political discourse. Code word usage can be indirect and complex. For political figures, being as abstract as talking about cutting taxes can serve as code for cutting social programs, which then is code for opposing welfare spending, which suggests the racialized images associated with welfare usage (Haney López 2014).

Code words have also broken into the public debate on the War on Drugs. Even those taking a presumably progressive stance in opposition to punitive drug policies and practices often employ code words to make their argument. And the use of code words in public discussions is not limited to print media. Racial code words are prevalent in online comments sections accompanying news stories (Hughey and Daniels 2013; Rosino and Hughey 2016). They are primarily employed to express political and racial meanings, yet avoid censorship (Hughey and Daniels 2013). The concept of code words can help explain why the terms like "inner-city," "urban," or "foreign" so commonly accompany descriptions of moral problems and position some groups as more threatening in the pages of the debate in newspapers all across the country.

Conclusions

The concepts of **racial silence, resonance,** and **code words** are indispensable tools for our conceptual toolkit. They are theoretical magnifying glasses that help bring findings on the War on Drugs debate into a crisp close-up view. They uncover some sociological dimensions of the debate that are less visible. They explain why certain frames (i.e., functionalist, fiscal, freedom, and justice) were prevalent and why others (racial unfairness) were uncommon in the arguments against the War on Drugs that appeared in newspapers since the early 1980s. For instance, racial silence could conceal the racial justice implications of the War on Drugs from a large segment of the American population. In the next chapter, I will examine the "unspoken conversations" (Zerubavel 2006) in an analysis of identity construction through engagement in these debates on the internet. These concepts will be helpful to return to as we continue our discussion about how people respond to claims about the War on Drugs through commenting on online news sites and specifically how commenters engage in identity construction through these responses.

These help us better understand **debate dynamics** or the overall patterns and themes in a public debate. In the context of these debate dynamics, why does identity matter? When individuals engage in discussions about contested social issues, they are also making broader claims about who they are, who others are, and their place in the broader social world. For example, statements within debates often include the term "we." "We" claims say a lot: who belongs or doesn't belong, who the speaker thinks they represent, or with what group or groups the speaker identifies. In the newspaper dataset, "we" claims ranged from suggesting the population of the United States, the government, or policymakers and excluding groups such as drug users, criminals, immigrants, or people living in the "inner-city."[7] These types of claims lay the ground for how people form identities through talking and thinking about contested social issues.

And where these types of claims especially proliferate is in spaces where not just those who can get their views published in a newspaper but almost any American with internet access can give their opinion such as the comment fields of online articles. So in the next chapter, we will see how audiences who consume news media about the War on Drugs respond with similar types of claims. Moreover, I will show how these claims provide opportunities for individuals to identify with **subject-positions** and form **symbolic boundaries** and how these positions and boundaries are racialized.

Discussion Questions

1 Can you think of any situations in your life in which **racial silence** is an expectation? What is it about these situations that would make talking about racial issues seem inappropriate? What might happen if a topic such as racism or racial inequality were broached in these settings?
2 What are some stories or ideas about how society works that resonate with you? Why? What are some stories or ideas about how society works that don't resonate with you? Why not? What is it about you (such as your worldview or your position in society) that influences what resonates with you or not?
3 Try to brainstorm as many code words as possible that you have heard that insinuate racial groups or are associated with race without explicitly naming racial groups or issues. What groups are these code words related to and why? Where did you encounter these code words? What do they imply about those groups?

Notes

1 And while some discussion of class occurred, there was also a routine lack of discussion of issues of gender in this debate. This is worth mentioning, considering that stereotypes and inequalities are shaped by structures and meanings of race, class, and gender (Collins 1993). Criminal stereotypes certainly take on gendered meanings and connect to ideas about masculinity and femininity. However, while gender silence on this issue is complicated and important, it is somewhat less surprising.
2 For instance, sociologist Ashley W. Doane (2003:563), noting a lack of explicit mention of race within a debate over the racial integration of West Hartford schools, wrote, "it served the objectives of both sides in the debate to downplay race" because "racial discourse has become so politically charged." Amanda E. Lewis (2003b:86), a sociologist,

in her research on racial issues in schools, found that "Whites resolutely avoid the subject. Others (primarily people of color) regularly talk about race or racism, but when they do so publicly, they pay for their expressions or 'transgressions' in feeling alienated, expending a lot of energy, being labeled, and having to contain their anger and frustration. So race becomes an issue constantly at play but only rarely named."
3 These included relations to chattel slavery, inheriting property, money, or a business, assistance paying for education or a home mortgage, government programs such as the Homestead Act or GI Bill, and help provided by acquaintances, friends, and relatives (Mueller 2017).
4 Drug Policy Alliance stands as one of the only major drug policy reform organizations that address racial justice issues in the War on Drugs.
5 See Dalton and Huang (2014) on "motivated forgetting."
6 While the reference to Muslims as Skittles seems out of place amongst these tech company references, it likely refers to an image shared online by one of the sons of Donald Trump that used the metaphor of a bowl of skittles in which some are poisoned to connect Syrian refugees to the threat of terrorist violence in yet another example in our long list of contemporary racialized moral panics (Sonnad 2016). Moreover, the idea behind using the tech references was apparently that companies such as Google could not possibly use an algorithm to detect and censor the names of their own companies (Miami Herald 2016).
7 I have, as you may be thinking to yourself in reading this, been making "we" claims throughout this book, sometimes suggesting that you as a reader and I as an author form some type of group or collective, or sometimes talking about all of society.

Chapter 5

Identity Construction in the Heat of Debate

A "public and cultural relation" has formed between the media content produced by the debate over the War on Drugs and the audience that consumes it (Schudson 1989:169). However, media technologies now blur the line between media production and audience consumption (Ross and Nightingale 2003). Digital technologies provide platforms that enable individuals to respond to events and issues covered by the news media. For example, the comment sections that accompany online news articles muddle distinctions between private conversations and public discussions. Internet comment sections are typically anonymous and allow people to have conversations that usually take place backstage.[1] At the same time, they are publicly accessible and open to engagement by a global audience.

In the United States, a **digital divide** that is class-linked and racialized continues to provide unequal internet access to residents who live in rural, poor, black, and some Latinx communities. Racial discrimination, age, region, income, education level, and generation all shape which Americans have reliable access to digital and online spaces (Robinson et al. 2015). Like the offline world, racial, ethnic, and class inequalities also characterize digital spaces. Inequalities in access to digital media platforms "play key roles in a range of outcomes, from academic performance to labor market success to entrepreneurship to health services uptake" (Robinson et al. 2015:570). Digital inequalities, like other forms of inequality, also influence the scope and inclusiveness of the public debates that increasingly unfold in these venues. Online spaces such as digital comment fields on news media sites are not merely ideal public spheres where people work together, propose rational arguments, and seek to find the common good (Daniels and Hughey 2013). Social forces such as ideologies, systemic forms of racism and sexism, and class inequality influence how people participate in online discussions.

In this chapter, I draw upon online comments on news articles to analyze how people participate in public debates in the digital sphere. In comment sections, people communicate responses to media content and other audience members, and in doing so, express, clarify, and construct identities (Rosino and Hughey 2016). First, let's dig a bit deeper into the relationship between media and audience, especially as it relates to racialized and contested social issues.

Media, Audiences, and Racial Oppression

Examining the production, distribution, consumption, and effects of media as continuous and connected processes helps us understand mass media's relationship with society (Hall 1980). The social context at each of these stages influences media content and its impact (Hall 1980). Because interactive media technologies blend the lines that separate these stages, we can also think of **digital media as a social practice** (Couldry 2012). In other words, people actively participate in media, and their ideas and identities shape this participation. Media scholars have attempted to answer "broad questions about the sorts of things that people are doing with media amid the proliferating complexity of digital media" (Couldry 2012:57).

Digital media platforms have reshaped the relationship of media to its **audiences** – the groups of people that witness, consume, and even participate in media events (Ross and Nightingale 2003). Media and communication scholars Karen Ross and Virginia Nightingale (2003:4) point out that media audiences are formed by "pre-existing social and cultural histories and conditions, and sometimes by a sense of shared interests that incline them to repeatedly use popular media vehicles." The development of media technology allows audiences to communicate with media content and each other. As Jay Rosen (2006), a journalism professor, wrote, "the people formerly known as the audience wish to inform media people of our existence, and of a shift in power that goes with the platform shift you've all heard about." These new possibilities for engagement mean that audiences can increasingly produce and share their interpretations of media content rather than acting as passive consumers (Ross and Nightingale 2003).

Media-audience relationships have long-held implications for racial inequality well before the digital media explosion. Robert E. Desrochers, Jr. (2002:623), a historian, wrote that "slavery and the newspaper grew up together in Massachusetts, in a close and synergetic relationship that made slave-for-sale advertisements a regular feature of the local press for most of the eighteenth century." People often think of the brutal system of racialized chattel enslavement in the United States as confined to the South. However, in New England as well, advertisements of enslaved people for sale were a standard fixture of early newspapers (Desrochers 2002).

These newspapers also routinely included runaway slave advertisements warning that enslaved people who escaped their bondage and sought liberation in these states may pretend to be free (Hodges and Brown 1994; Marshall 2010). As these slave ads demonstrate, during the birth of mass media in the United States, these new forms of communication helped their producers and audiences facilitate racial oppression.

With increased access to printing presses and education came the growth and diversification of print media producers and audiences. In 1827, the first newspaper written by and for black citizens, *Freedom's Journal*, was created in New York City (Pride and Wilson 1997). The ongoing struggle for racial equality, citizenship, and full rights motivated the emergence of the black press and the need for black journalists to express the concerns of their communities (Pride and Wilson 1997). While this work was crucial to the fight for liberation, black journalists and editors often faced backlash for this cause.[2]

Ida B. Wells, a journalist and sociologist, edited the Memphis, Tennessee-based *Free Speech and Headlight*. The paper reported on the racial terrorism of lynching in the early 1890s. While Wells went on to become a significant activist and journalist, in 1892, a mob of whites, outraged by this reporting, wrecked the *Free Speech*'s printing press and offices (Pride and Wilson 1997). Magazines, radio, and television also became venues for black-operated media. In 1920, sociologist W.E.B. Du Bois founded the highly successful magazine of the NAACP, *The Crisis*. As founding editor, Du Bois "wrote editorials of force, beauty, cold irony, and sharp thrust" (Pride and Wilson 1997:251) on a range of issues impacting black Americans.

The black press provided a crucial means of expressing the conditions of black citizens and a space for critical narratives on racialized issues and events. Consider how newspapers reported a violent conflict between a group of black and white residents in Detroit in the 1940s. In what's now known as the 1943 Detroit race riot, whites heavily outnumbered blacks, and blacks suffered the brunt of the property damage, injuries, and deaths that ensued (Sitkoff 1969). Detroit's black-owned newspapers such as *Michigan Chronicle* provided relevant commentary and reporting on racial terrorism by the Ku Klux Klan, housing shortages, and economic exploitation of black workers (Kapell 2009). However, white-run local newspapers such as *The Detroit News* or *The Detroit Free Times* only presented the perspectives of white residents and officials. They perpetuated a narrative that black "hoodlums" had been the primary catalyst (Kapell 2009). For the white reader who learned about these events through reading *The Detroit News*, this media content played a vital role in shaping their understandings of what occurred, and legitimating the violence against blacks and more broadly of the status quo in Detroit.

Media content provides a source of **social information** for audiences (Entman and Rojecki 2001). People use the messages they receive from media to form opinions on social issues, a sense of who they are, and ideas about their place in the social world. The views, images, and narratives that appear in media content shape, if only in indirect ways, what those who consume them define as real or true (Dixon 2007). But not everyone responds similarly as an audience member. Depending on their background, worldview, and position in society, audiences may derive alternate or critical interpretations, intended interpretations, or even a mix of both when they consume media (Hall 1980). For example, we can imagine that black audience members may have derived a more critical interpretation from mainstream white newspaper coverage of the 1943 Detroit race riot. It likely clashed with their lived experiences of racial oppression or their understandings of the events' causes and consequences.

When we talk about racial segregation in residential patterns and social life, we often focus on the impacts on black, Latinx, Asian origin, or Native people. But as noted by sociologists Eduardo Bonilla-Silva and David G. Embrick (2007), the preferences, opportunities, and worldviews of whites are shaped by their high likelihood of living in racially homogenized environments. Whites who are isolated from people of color in their neighborhoods, schools, everyday interactions, and close relationships are much more likely to rely on media content as a source of information on racial issues (Entman and Rojecki 2001; Mastro and Tropp 2004). Because media images and narratives can stand in for real experiences, media content that includes racial stereotypes is often used by whites to justify racial inequality. For instance, a study on how news media affects audiences conducted by media scholars Robert M. Entman and Andrew Rojecki (2001:9) concluded that "the racial stereotyping of Blacks encouraged by the images and implicit comparisons to Whites on local news reduces the latter's empathy and heightens animosity."

Even though they may seem disconnected, there is a robust relationship between media content, audiences, and the continuation of racial oppression. And the relationship between new media content such as online news and the way that audiences make meaning of that media content has essential consequences for contemporary society. But what happens when audiences are not just consuming media but also participating in collective processes of definition and interpretation? Within online spaces that allow people to provide commentary, are audiences just making sense of media content when they respond to it? Perhaps they are also making sense of themselves, others, and the world around them.

Commenting and Constructing Identities

Media-audience relationships are made more complicated by new media technologies (Ross and Nightingale 2003). New online platforms and interfaces enable people to share commentary on media content. Comment sections allow people to respond to issues and events presented in news and opinion pieces. They are arenas where people engage in heated exchanges over the definition of contested social issues. Especially on major news sites, comment sections can be extremely active and interactive.

These digital spaces allow people to express and define themselves through communication about public events and issues and their presentation in the media. So, they can help us understand media audiences. And as anyone who has spent substantial time on the Internet knows, they are a pervasive feature of contemporary digital media. In an April 18, 2006, edition of "The Opinion Pages" section of *The New York Times*, a series of op-eds discussed the issue of comment sections on news articles and whether they facilitate public discussion on important issues. Moreover, in a particularly "meta" example of the commentary function of contemporary media, each of these op-eds was accompanied by a comment section where people articulated their perspectives on comment sections. But what can analyzing these spaces of communication and reaction help us understand about things like racial identity, media, politics, and power?

In 2008, a fabricated controversy emerged over the birthplace and citizenship of Barack Obama. Despite substantial evidence to the contrary, conservative politicians, pundits, and activists questioned his US citizenship throughout his 2008 campaign and during the first term of his presidency (Zernike 2011; Parlett 2014; Dropp and Nyhan 2016). Businessman and television celebrity turned President Donald J. Trump used this narrative to help launch his political career (Barbaro 2016). And commenting on news stories about the "birther" phenomenon allowed people to make racial claims about who does and doesn't belong within the bounds of citizenship (Hughey 2012b). In talking about racialized events and issues, people make claims about the meaning and relevance of racial categories in relation to those events. In doing so, they construct and communicate their racial identities.

In the first chapter of this book, I mentioned **four aspects of identity**:

1 Identity is an ongoing process rather than something fixed.
2 Identity is constructed, in part, through language.
3 Identity is shaped by choices people make and how society is organized.
4 Identity categories reflect a sense of social group differentiation, inclusion, and exclusion.

These propositions may seem somewhat intuitive when we consider our own sense of self in everyday life. Perhaps more importantly, though, they help us answer some big questions about the relationship between individuals and society.

We previously discussed how the "definitions of the powerful" (Hall et al. 1978:57) influence media content and that it, therefore, reflects dominant sets of ideas or *ideology* (Hall 1995). But what about identity? For one thing, when these dominant ideas gain our attention, we have an opportunity to identify ourselves by conforming our thoughts and actions accordingly (Althusser [1971]2001).[3] If you tend to encounter the idea that men are rational and you identify as a man, you may be tempted to identify with that idea ("as a man, I am rational"). You may then act and think in ways that reflect that idea (like avoiding "irrational" activities such as expressing emotions) (see Connell 1995). This **process of identification** allows you to define yourself. It also impacts how you treat others, how you present yourself, and how you think and feel.

Similarly, responding to media content about the War on Drugs provides opportunities for people to identify themselves (and others), and this process of identity construction has considerable social consequences. Undoubtedly, we all bring a previously established sense of self to conversations about social problems. We may have already developed a perspective on the issues, ourselves, and society well in advance. However, identity construction is not just a one-time event. People's identities are often refined and clarified during conversations about social issues (Walsh 2004). People engage in this ongoing process of defining themselves, others, and contested social issues while reacting to media content and the comments of fellow audience members.

People often make sense of their experiences through reference to "the self" (Mead 1934).[4] For instance, someone who feels remorse might say, "I just can't forgive myself." Sociologist Erving Goffman (1959, 1967) built on this idea by arguing that "the self" is not something natural or fixed. Instead, he pointed out that people form a sense of self (or identity) through communication, often intending to leave an impression on others.[5] In other words, people strategically produce their identities because they have "social and material consequences" (Howard 2000:371). People define their "selves" while interacting with others in everyday life to achieve individual goals, make specific claims, or attempt to be perceived by others in a particular light.

We use words and language to develop a sense of who we are. And we attempt to define who we are, or engage in identity construction, by being strategic about how we interact with others. But our identities are not just products of our individual choices and strategies. The way the society around us is organized and our social position within it also matters. Sociologists Peter L. Berger and Thomas Luckmann (1966:173) wrote:

> Identity is formed through a social process. Once crystalized, it is maintained, modified, or even reshaped by social relations. The social processes involved in both the formation and the maintenance of identity are shaped by the social structure.

So how does the social structure influence identity construction? Consider the racial categories that people use to identify themselves and others every day. As we discussed in Chapter 4, social and political processes create and change racial categories. These processes do not just reflect individual decisions or small-scale social interactions. They are products of group relations of conflict, solidarity, and oppression as well as laws, economic systems, political and social institutions, and forms of cultural representation (Omi and Winant 1994; Hall 1997; Haney López 2006).

We do not construct identities from scratch. We use categories and narratives spread by institutions such as mass media, the state, and the education system, and even friends and family.[6] When these categories and stories come together to form a reasonably clear picture of the world and how it works, they make up **discourses**.[7] Discourses are the resources that people have at their disposal to make sense of themselves, other people, the world around them, and their place in it. They are also resources that people draw on in the "process of collective definition" (Blumer 1971) that influences how we approach social problems like the War on Drugs.

For example, while some people agree that racism is wrong, how do they define racism? In 2016, the definition for "racism" in the Merriam Webster dictionary website was polysemic. In other words, the word has several potential meanings. These include first a "belief," second "a doctrine or political program" or "political or social system," and finally a form of "prejudice or discrimination" (Merriam Webster 2016). Likewise, different discourses present different ways of defining things. People using the dominant racial discourse see racism as individually held attitudes about racial groups – and thus claim that anyone, regardless of their position in society, is engaging in racism if they act or think in an intentionally racially prejudicial way (Doane 2006). It is perhaps unsurprising in this context then that increased awareness about the systemic and institutional aspects of racism has led to efforts to update and refine this definition (Hauser 2020).

College student and activist Kennedy Mitchum, a leader in pushing for these changes, realized the consequences of these racial discourses during a debate about racism. In this conversation, a white classmate "copied and pasted only the first section of the Merriam-Webster entry [which defines it as a belief] without recognizing the second part and its wider implications in society" (Hauser 2020). In contrast, those employing a more critical racial discourse see racism as systemic or connected to the overall structure of society, perhaps focusing on racially unequal patterns in how institutions and people dole out rewards and punishments (Doane 2006). Definitions and the relationships people hold between ideas or concepts have significant consequences for their social and political actions (Hall 1995). So, these racial discourses generate conflicting answers to whether and how we should address racism as a society.

Constructing your identity means taking on a perspective. From this perspective, you view and define the world around you and your place in it. When people tell stories and evoke categories to communicate a stance or viewpoint while debating contested social issues, these discourses create **subject-positions** from which problems and events make sense to them (Hall 1997). We form a sense of who we are when we identify with these subject-positions and take them up as our own (Hall 1996, 1997). We then see "the world from the vantage point of that position and in terms of the particular images, metaphors, storylines, and concepts which are made relevant" (Davies and Harré 1990:46).

Our sense of self also depends on how we think about social categories or groups. We construct identities through categorizing things as different from ourselves and one another. As strange as it sounds, it is because you have an idea of what is "not you" (such as other people and objects) that you have any idea of who you are.[8] As Stuart Hall (1996:5) argued, "identities can function as points of identification and attachment only because of their capacity to exclude, to leave out, to render 'outside,' abjected." Sociologists often study how people create **symbolic boundaries** – "conceptual distinctions" that allow them to "categorize objects, people, practices, and even space and time" (Lamont and Molnár 2002:168). A symbolic boundary between "us" and "them," for instance, allows people to identify themselves as part of a group and exclude others.

Distinctions between racial categories such as "black" and "white" are symbolic boundaries that "separate people into groups and generate feelings of similarity and group membership" (Lamont and Molnár 2002:168). Although myths about the genetic, scientific, or biological basis of race persist, these categories don't reference anything real in the sense of biological or essential and immutable human traits (Graves 2001).[9] They are the products of history and shaped by institutions, group interests, and conflicts (Omi and Winant 2014).

Racial categories and the dominant meanings attached to them, nonetheless, have real effects on society due to the ways that people use them in everyday life (Bonilla-Silva 1997, 1999; Lewis 2003a). Public debates shape the meanings of racial categories and their differences. But they only have a significant influence on social outcomes like racial inequality when they become prevalent (Lamont and Molnár 2002; Doering 2014; Omi and Winant 2014). And dominant groups, such as those categorized as white, tend to have more collective power to have their interpretations of racial categories and differences influence the prevailing "common sense" about racial issues (Feagin 2010; Rosino 2016).

How we define members of racial groups and make distinctions between them determines how we perceive and act upon the racialized inequalities in society. So, we can't just focus on the production of racial categories on a large scale. We also need to consider how people think and talk about them in daily life. For example, claims about morality and cultural ideals play a vital role in what "being white" means to many white people (Lamont 2000). White people construct symbolic boundaries between not only whites and nonwhites, but also between "good" and "bad" whites by connecting what it means to be white to essential and ideal traits (Lewis 2004; Hughey 2012a).

Debates over contested social issues can act as battlegrounds of identity (Doering 2014). They often seem like contentious arguments that produce more heat than light due to the high amount of friction involved. In this sense, discussing contested social issues is not merely about gathering information or solving problems by bringing issues to light but also forging a sense of who "we" are in the fires of group conflict. So, let's look at the patterns within comments on media content on the War on Drugs.

Analyzing Online Comments

It would be impossible to talk with the audiences of our newspaper manuscripts without millions of dollars and perhaps a time machine. So, I collected data from a source that can provide information about how audiences respond to claims about the War on Drugs in the media – online comments.[10][11] Three thousand one hundred forty-five internet comments on articles on the War on Drugs from 2009 to 2014 were analyzed to understand how people formed identities through their engagement with this debate. Let's discuss the frames and themes from the comments (Table 5.1) and how they relate to the processes of identity construction.

Table 5.1 Frames and Themes in Comments

Frame/Theme	N	%
Frame: Fiscal	**157**	**0.0860**
Opportunity cost	35	0.2229
Expense	122	0.7771
Frame: Freedom and Justice	**480**	**0.2629**
Class	40	0.0833
Police militarization	25	0.0521
Hypocrisy	31	0.0646
Corruption	170	0.3542
Mass incarceration/overcrowding	24	0.0500
Civil liberties	162	0.3375
Constitution	30	0.0625
Harms individuals/families	29	0.0604
Frame: Functionalism	**525**	**0.2875**
Failed/unwinnable	56	0.1067
Doesn't reduce drugs	59	0.1124
Treatment	34	0.0648
Regulation	117	0.2229
Black market/criminal/violent/other	132	0.2514
Drugs are harmless/beneficial	127	0.2419
Frame: Racial Unfairness	**219**	**0.1199**
Motivated by racism	31	0.1416
Structural racism	63	0.2877
Racial bias	102	0.4658
Black communities susceptible	23	0.1050
Frame: Racialized Victim Blaming	**278**	**0.1522**
Criminality	126	0.4532
Pathology	96	0.3453
Denial of racism	56	0.2014
Frame: Colorblind Victim Blaming	**116**	**0.0635**
Punish criminals	116	1.0000
Frame: Colorblind Defense	**56**	**0.0307**
Anti-drug	33	0.5893
Legalization would fail	23	0.0126
Total	**1,826**	**100.00**

Familiar Frames

Many of the frames and themes within these comments mirrored those that emerged in our previous examination of newspapers. I will cover these frames briefly. However, there were also frames and themes unique to internet comments. We'll also talk about how commenters formed **symbolic boundaries** by talking about "in-groups" and "out-groups." And we'll discuss how they identified with **subject-positions** in the discourses they used to make claims. First, let's look through how audiences articulated familiar frames from Chapter 3, such as the fiscal frame.

The Fiscal Frame

Much like in newspapers, commenters focused on framing the War on Drugs as a financial issue. A significant theme was the economic opportunity costs of the War on Drugs. One commenter asked, "why are we missing the opportunity to tax this and bring in much needed revenue to the states and federal government?" (2012, *Forbes*)[12]. Commenters also decried the War on Drugs as a waste of taxpayer money. Many of these claims hinged on ideas about how the War on Drugs conflicts with ideals of fiscal conservativism or fiscal responsibility. One commenter pointed to this contradiction by noting that it is "Time for 'conservatives' to think about just how 'conservative' it is to support a huge government spending program called the 'war on drugs.'" (October 20, 2014, *Daily Signal*). Others used statistics to demonstrate the enormity of the spending: "Over four decades [...] American taxpayers have dished out $1 trillion on the drug war" (April 8, 2013, *Huffington Post*). Claims within this theme also often contained references to history, as one commenter humorously noted, "those are your tax dollars being flushed down the crapper of Richard Nixon's paranoid delusions" (October 13, 2014, *Washington Post*).

The Freedom and Justice Frame

Within the freedom and justice frame, commenters argued against the War on Drugs by critiquing the use of militarized police as a means of social control. One comment writer contended, "The War on Drugs isn't primarily about drugs, it's a medium for social repression and the advancement of the national security/police state" (October 18, 2014, *Washington Post*). Others pointed to the example of police responses to demonstrations against anti-black violence and police brutality:

> The "war on drugs" has been the primary driver of militarization of police forces in the U.S. Had the Ferguson PD not had all that **military** equipment, they might have made a less aggressive and less extreme response to the initial peaceful protests and the riots might never have happened at all.
> (October 14, 2014, *Washington Post*)

Claims against the War on Drugs in comments that did not appear significantly in newspapers pointed out that its supporters are hypocritical. This theme criticized groups that claim the moral high ground or whose ideology seems to clash with the implications of the War on Drugs such as conservatives, Christians, and politicians in general. One person responded to others defending the War on Drugs as a way of dealing with immoral drug use by quipping:

> Hard-asses that want punishments to far exceed a sensibl [sp] punishment for a crime, kissing the butts of any politician that claims to be "tough on crime!," etc. Leading the charge when it comes to abridging people's rights to the privacy of their own homes, or even to the famous "Life, Liberty and the Pursuit of Happiness." Then, suddenly, it's them or their kid in the clutches of the criminal justice system. Such shock! Such anguish! Such alligator tears! Wailing "how could this be???"
> (April 8, 2013, *Huffington Post*)

Commenters also criticized the hypocrisy of laws targeting less severe crimes, while more serious elite crimes such as financial fraud and corruption go unpunished. One comment stated, "In addition to the numbers and the cost to the people and their families and communities this targeting of marijuana [...] at the same time as the war criminals and all the Banksters went free is an abomination in the land of the free" (October 18, 2014, *Washington Post*). And in an outgrowth of libertarian arguments against the War on Drugs, comment writers also argued that it violated the constitution or states' rights: "Care to show me where the federal government has the authority to ban drugs? They needed an amendment for alcohol... why not for drugs? Ah, right, they just ignore the Constitution" (April 8, 2013, *Huffington Post*).

Mass incarceration/overcrowding was another prominent theme. Appeals to statistics and personal experience were common. Someone shared the following anecdote:

> I once worked in a security level 2 federal prison. Over half of the inmates there were in for drugs and ONLY drugs. [...] That prison was holding 275% of the prisoners it was designed to hold [...]. Even worse, the prison was designed and built for WOMEN prisoners, but it holds only MALE prisoners....
> (April 9, 2013, *Huffington Post*)

Commenters also made international or historical comparisons. One person asked, "does it bother you that we have a higher percentage of our own people locked up than any other country every has ever had in the entire history of the human race?" (2012, *Forbes*).

Another theme focused on the prison-industrial complex. The majority of these comments implied cooperation from a multitude of actors, including politicians and government, pharmaceutical corporations, the financial industry, the wealthy, private prison corporations, lobbyists, and even drug cartels. One comment writer on a story about the racial disparities caused by the War on Drugs responded: "This story is yet more evidence that the 'drug war' bureaucracy including law enforcement, judges, lawyers, guards, prison administration, etc. just don't get it. Their goal is to grow their precious bureaucracy, period" (April 8, 2013, *Huffington Post*). Civil liberties and general concerns that the War on Drugs harms individuals/families were other concerns for many commenters. One comment pointed out "the human toll, broken families, broken individuals, eroded civil rights" (April 8, 2013, *Huffington Post*) caused by the War on Drugs.

Finally, comments emphasized that the War on Drugs was an issue of economic class inequality. One commenter wrote, "whole neighborhoods of the poor have been decimated" by drug law enforcement. Interestingly, people's comments enforced racial silence through the logic that economic class inequalities were the real matter at hand rather than racial oppression. In other words, commenters also brought up social class to discredit claims that the War on Drugs was primarily a matter of racial inequality. One comment pointed to the relationship between social class and the ability to avoid criminal sanctions. It argued that among the poor, "the drug war is being used to harm Americans of all races, creeds, religions, and political bent" (October 8, 2014, *Washington Post*). Given the relationship between racial inequality and class, economic inequality certainly plays a significant role in who is impacted negatively by the War on Drugs. Yet, the overall impact of this line of reasoning is to discredit the clear evidence that the War on Drugs is a deeply racialized issue.

The Functionalist Frame

As we saw in Chapter 3, the **functionalist frame** is all about how the War on Drugs does not function in achieving the goals of creating a lawful and orderly society and reducing crime and drug use. Commenters argued that the War on Drugs doesn't decrease drug use or the availability of drugs. One person remarked in a comment, "Right now, you could drop me down in any city (even medium sized town) and I could find just about any illegal drug that exists" (2012, *Forbes*).

Comment writers also argued that legalization would be more functional. Several claims within this theme presented legalization as a commonsensical, logical, or rational alternative. These claims also distinguished between drugs and the types of drug use worthy of legalization, such as cannabis and those that were not worth the risk. Comments also appealed to historical examples, as one commenter argued, "Judging from the article it would be prudent to take the advice given to then President Nixon and remove Marijuana from the list of scheduled drugs completely" (October 8, 2014, *Huffington Post*). Alongside legalization, another functional alternative posed was treatment. While well intentioned, these comments tended to ignore the racialized and classed dimensions of the current approach of treatment. As we saw in Chapter 2, empathy for drug addicts as facing a difficult brain disease rather than moral failings, or as troubled individuals rather than threats to society is already in place, yet remains reserved for white, upper-class drug users.

A theme common among comments but not newspapers was redefining drugs as neutral or positive. Many made comparisons to alcohol: "I also find it ironic how the government can allow something like alcohol to be legal which kills 2.5 million people a year. Marijuana kills 0. because it is not toxic" (2012, *Forbes*). And others argued using anecdotal evidence that cannabis was not a cause of crime or violence: "Never heard of a domestic abuse case or a bar fight caused by the guy smoking pot" (2012, *Forbes*). Commenters also posed personal experience to argue that cannabis provides benefits or does not create dysfunctional individuals. One commenter noted their cannabis use and identified as a "straight A student" who is "about to graduate with my masters" (2012, *Forbes*). Another praised the "holistic pain relief" (2012, *Forbes*) it provided.

As we discussed in the second chapter of this book, increasingly, the dominant narrative pushed by organizations and activists that have sought to legalize cannabis has centered upon white middle-class users who partake in the drug in responsible recreational ways or for medical uses (Provine 2007). This approach to criticism of the War on Drugs has been useful for gaining empathy from the general public for the unfairness or impacts of cannabis prohibition on people accepted as having potential or promise without addressing the real racialized power dynamics involved.

Another theme was the sense that the War on Drugs is a failure. One commenter remarked, "like most wars America has declared … this one was lost from the very beginning" (April 8, 2013, *Washington Post*). Pleas for sanity often accompanied claims in this theme: "nobody sane would ever support [the War on Drugs], after carefully reviewing all the evidence" (2013, *Forbes*). Another argued, "if you support [the War on Drugs] you must be either ignorant, stupid, brainwashed, corrupt or criminally insane" (2012, *Forbes*). Comparisons to alcohol prohibition were also frequent: "Remember Prohibition? Go to jail for selling, or possessing alcohol ?? That didn't work either" (April 8, 2013, *Washington Post*) or "Have they never heard of Al Capone? Or any one of the numerous thugs who became MILLIONAIRES from a little policy called PROHIBITION? Is this not clear history?" (2012, *Forbes*).

Similar to the newspaper dataset, a large portion of comments in the functionalist frame argued that the War on Drugs was dysfunctional because it created criminals. Often these claims took on racialized language and identified criminals with racial groups. Commenters used the idea of a **racialized criminal threat** to make sense of the dysfunctionality of the War on Drugs. Commenters often referred to the dangerous individuals encapsulated by this idea as "they" or "them": "we put them in little cages and wonder why they want to kill each other" (April 8, 2013, *Washington Post*). When specifically naming this subject, commenters argued that the War on Drugs was empowering and enriching "thugs and criminal empires," "common thugs," "illegal invaders," "the Taliban and the terrorists of al Qaeda," and "terrorists as well as drug cartels."

One commenter argued that the War on Drugs helped "evolve local gangs into transnational enterprises with intricate power structures that reach into every corner of society, controlling vast swaths of territory with significant social and military resources at their disposal" (2012, *Forbes*). Images of racialized threat also involved implications of nonwhite racial groups colluding to upend the social order: "With social demographics changing within the country this will lead to serious civil unrest. The Muslim extremists are recruiting Latino and black men" (October 4, 2014, *Huffington Post*).

Moreover, commenters critiqued the assumed naivety of those who denied the supposed "harsh truth" about racialized criminals: "Come out of your ivory tower and get down to crack alley" (2012, *Forbes*). Ironically, considering the role of moral panics in the creation of the "War on Drugs," many of these categories, narratives, and ideas fall under the rubric of racialized moral panics. In doing so, they draw on the specter of racial threat – echoing those historical examples we discussed in previous chapters such as the "yellow menace" or the "Negro cocaine fiend." The racialized claims within this theme argue that the War on Drugs is problematic not because it harms people of color, but because it fails to protect "us" from "them."

The Racial Unfairness Frame

The **racial unfairness frame** in comments communicated that the policies and practices of the War on Drugs contribute to unequal outcomes between racial groups in the United States. For instance, commenters often discussed racial bias: "Regardless of anyone's opinions on this topic, can we at least agree that the racial numbers (1 out of 15 blacks versus 1 out of 106 whites incarcerated) tell us that something is definitely wrong with this picture?" (April 8, 2013, *Washington Post*). Within this theme, a symbolic boundary emerged between an out-group of police and whites and an in-group of blacks. One commenter simply stated, "Great nation, when you're white" (October 4, 2014, *Huffington Post*).

Commenters often identified themselves as nonwhite or shared their personal experiences with the unfairness of the War on Drugs:

> Whites don't understand getting arrested for nothing beacause [sp] it simply doesn't happen to them. I'm a black university student nerd, who wouldn't commit any sort of crime, but I get questioned and cuffed way before my white or asian friends.
> (October 5, 2013, *Huffington Post*)

Finally, commenters argued that claims that racism and discrimination are not salient issues are naive or unrealistic: "Unfortunately, for one group of people, this is pure hyperbole while for another group of people this is a harsh reality. 'Race-baiters' are not the problem. It's being naive that hinders us from becoming a post-racial society" (October 5, 2013, *Huffington Post*).

Some commenters also stated in clear terms that the War on Drugs is motivated by racism against people of color. One argued that "Marijuana was made illegal because of racism" (April 8, 2013, *Washington Post*).[13] Another stated, "Sounds like a war on black youth, not a war on drugs" (April 15, 2013, *Washington Post*). Others pointed to historical evidence linking drug policies to racial oppression: "When the 1937 Marihuana tax act appeared it was aimed directly at latino and black people right at a time when they were 'rising above their station' to become famous musicians and sportsman" (2011, *Forbes*).

Comments also presented racial unfairness as a product of black communities having a unique susceptibility to drugs and crime. One commenter asserted that "Given that crack is sold mostly in the ghetto – although often to white users – the result of the policy was to flood federal prisons with young black men" (April 8, 2013, *Washington Post*). Another argued that racial unfairness in the War on Drugs was caused by "the way drugs are sold in black areas open in the streets while in the white and other areas it's done in secret behind closed doors" (April 8, 2013, *Washington Post*).

Finally, commenters often argued that the War on Drugs is part of how society is arranged to perpetuate racial inequality. This theme reflects the concept of **structural racism** – that racial inequality is a product of social structure or the organization of society in terms of rules and the distribution of resources. One commenter contended that the War on Drugs "was designed to do exactly what it did, incarcerate as many black males as possible" (April 8, 2013,

Washington Post). Moreover, likely influenced by discourses circulated by books and documentaries on the topic, comments noted similarities or continuity between the War on Drugs and racialized mass incarceration and historic forms of racial oppression. One comment writer stated, "the white man's 'War On Drugs' is merely an extension of the white man's 'War On Non-White's' which the white man began in October of 1492" (October 8, 2013, *Huffington Post*). Another argued, "Prisons are the new plantations…" (October 8, 2013, *Huffington Post*), and another posed the following equation: "For-profit prisons + racial profiling + availability of unbelievably cheap and forcibly submissive prison labor to big businesses = the new slavery" (October 4, 2013, *Huffington Post*).

Unique Frames: Othering, Boundaries, and Identity

Comments on articles on the War on Drugs in online news sites also reveal the role of racialized symbolic boundaries (cf. Lamont 2000; Hughey 2012a) in how people form identities through making claims about contested social issues. Distinctions between racial groups that implied essential traits were used to construct identities within the War on Drugs debate. For example, associations between racial groups and ideas of security and threat often accompanied such group distinctions. This imagery, depicting racial groups as threatening or dangerous, was present in justifications for both dismantling and preserving the War on Drugs. One commenter imagined the following scenario post-War on Drugs:

> Gangs would be out of business, little street corner punks would be out of business, courts would be freed up to prosecute "real" crime and prison costs would be less than what we are paying now to house these drug prisoners.
> (April 8, 2013, *Washington Post*)

While another commenter, arguing why the War on Drugs must continue, claimed:

> Dealers are violent criminals by nature. They aren't otherwise innocent people. They don't care who overdoses and dies on their crack. Their business is the reason that domestic terrorist street gangs infest our city centers.
> (April 8, 2013, *Washington Post*)

Though many of the online comments mirrored the frames within newspapers, unique and even overtly racial frames were employed within comments, both criticizing and defending the War on Drugs. Examining these frames helps us understand a broader range of how people talk and think about this issue. It demonstrates what types of ideas and narratives audiences employ to understand the debate that takes place via mass media, and how people form identities through their participation.

Racialized Victim-Blaming Frame

A significant frame within comment sections but not the debate in newspapers on the War on Drugs (even among its proponents) was **racialized victim-blaming**. William J. Ryan, a psychologist and sociologist, defined victim-blaming as "justifying inequality by finding defects in the victims of inequality" (Ryan 1976:xiii). And in the debate over the War on Drugs, the act of victim-blaming took on racial meanings. It connected to racialized symbolic boundaries or the distinctions that commenters made between black and white. Blackness was often equated with criminality in comment sections, primarily in response to claims about the racial unfairness of the War on Drugs.

These claims interpreted racial disparities in arrest or mass incarceration as a natural or legitimate outcome of inherent differences in traits between whites and blacks. Despite evidence that whites are just as likely, if not more, to commit drug crimes (Fellner 2009), the trope of the black drug dealer or street criminal was commonly evoked. One commenter insinuated the high rate of drug-related incarceration among blacks was justified by arguing, "Lots of Black People Sell'n Dope, Lots of Black People In Prison" (April 8, 2013, *Huffington Post*).

Claims about blacks as inherently criminal frequently included the use of "the" before the name of a racial group. One commenter wrote, "Quit the liberal hype! [...] They forgot to mention that *the* blacks are more likely to be the ones selling the drugs. More blind reverse discrimination" (April 8, 2013, *Huffington Post*). This code word allows people to engage in what philosophers and social scientists call **othering**. Even something as subtle as the word "the" can reinforce symbolic boundaries. Linguist Lynne Murphey (2016) wrote, "'the' makes the group seem like it's a large, uniform mass, rather than a diverse group of individuals. This is the key to 'othering:' treating people from another group as less human than one's own group."

Audience members often made pains to explain away racially unequal outcomes in the War on Drugs as caused by anything other than systemic racism. This theme generally constructed a racialized sense of "us" and "them" – an out-group of blacks and, to a lesser extent, Latinx and an in-group of whites. Claims also depicted those criticizing racial injustice as making invalid claims or lying. Many responses pointed to "truth," "reality," or "facts" to make their counterclaims. One person wrote, "the fact of the matter is, and the article 'discourages' us from paying attention to that cold, hard fact, but blacks and Latinos commit a higher proportion of crimes than whites" (October 4, 2013, *Huffington Post*). However, reported crime rates are not objective measures of the amount of lawbreaking that takes place amongst a group of people. Reported crime statistics are skewed by whether the criminal legal system targets certain groups and communities, therefore, making them more likely to be arrested and charged (Hall et al. 1970). Moreover, disparities in reported rates of crime are much smaller than the tremendous rate at which blacks and Latinx are disproportionately imprisoned, stigmatized, and targeted by the criminal justice system in comparison to whites (Western 2007; Armaline 2011).

This theme provides one of the most striking examples of how people refine and communicate their sense of racial identity through participation in the War on Drugs debate. Comments overtly distinguished between "whites" and "blacks," and tied these categories to a distinction between "us" and "them." In doing so, commenters not only identified with the subject-position of "white" sharply juxtaposed with "black," but also specific notions of what it means to be white or black. These claims relied on categories and narratives from a dominant and longstanding discourse of criminality as a "racial trait" that, as we saw in Chapter 2, emerged during debates over race and crime at the turn of the 20th century.

Many claims within this theme attacked not only the notion that the War on Drugs unfairly targets black and Latinx people but also the related insinuation that whites are just as likely, if not more so, to commit drug crimes. A multitude of comments amplified the assumed criminality of blackness through contrast with the presumed innocence of whiteness. One commenter argued:

> white kids are not more likely to become drug users... That is the liberal media pushing their jungle fever pitch on the public.... There is a reason why 50% of the prison is blacks. Cops

and state attorneys have to prove cases, not simply say this guy is guilty because he is black… The antiquated white guilt of the boomers is really really old…
(April 8, 2013, *Huffington Post*)

Another posed the following rhetorical questions: "How many white kids you know selling drugs out front of a liquor store? How many white kids are smuggling drugs across the border?" (July 24, 2013, *Huffington Post*). Symbolic boundaries between categories such as white/black and innocent/criminal often overlapped so that white stood in for innocent and criminal stood in for black. One commenter, for example, argued:

Criminals tend not to understand the concept of delayed gratification and look for easy money. I get a kick out of people who make fun of white people because we're boring or stiff and take school seriously, go to college and take that seriously, have two parents raising kids, obey the law, know the law, etc.
(October 5, 2013, *Huffington Post*)

Beyond a focus on criminality, audience members also made remarks that equated blackness with pathology and dysfunction in general and related to things like family, behavior, and values. This theme reflects a "widespread narrative that black and Hispanic people possess an array of dysfunctional traits" (Hughey 2012:62). Many comments expressed **cultural racism**, whereby racially unequal outcomes are explained as solely the product of the cultural traits of people of color (see Bonilla-Silva 2014). This theme echoes 19th-century ideas known as Social Darwinism, whereby the logic of the "survival of the fittest" was erroneously applied to society to explain inequality. Yet, as noted by sociologist Stephen Steinberg (1989:79), "notions of biological superiority and inferiority have been replaced with a new set of ideas that amount to claims of cultural superiority and inferiority." Within this theme, comment writers depicted black families as dysfunctional, often accompanied by uncited and inconsistent statistics:

75% of the black men in federal prisons come from single parent homes mainly headed by women. Solve this problem and many of our social ills will resolve. Young boys need a strong hand in these critical years. Some woman [sp] can do this most cannot. These boys need fathers.

(April 8, 2013, *Huffington Post*)

What do they expect when only 15% of African American mothers have children within wedlock? You'd think they'd give a damned about having a father around before jumping in the sac [sp].

(October 5, 2013, *Huffington Post*)

The logic behind these claims reflects backward, yet tragically common, understandings of the relationship between families and societies. Differences in families' abilities to provide resources and opportunities to their members can help maintain social inequality (Bourdieu 1998; Lareau 2011). However, families do not exist in a void. They are embedded within their surrounding communities, institutions, and society at large. In the late 19th century, W.E.B. Du Bois developed a holistic approach to black families that revealed the impact of social forces like racial oppression. Du Bois (1889) examined the relationship between families and the social and material conditions of their communities that serve as an external source of resources and constraints. This approach demonstrates poverty, racism, mass incarceration, discrimination in employment, residential segregation, and public policies shape racial patterns in family formation (Du Bois 1889; Steinberg 1989; Clear 2007). Indeed, the War on Drugs itself has multiple harmful impacts on the economic status and stability of black families (Levy-Pounds 2010).

Many claims within this theme were much broader, with some including detailed lists of all the ways that commenters perceived black families, communities, and individuals as pathological. We would be remiss to merely chalk such claims up to ignorance or intolerance without recognizing that they also reinforce and strengthen dominant forms of white racial identity construction. Categorizing large and diverse racial groups as fundamentally deviant or morally wrong is based on arbitrarily selected criteria that enable dominant group members to define themselves as normal and moral.

A commenter described the racial disparity in the War on Drugs as a result of "disparity in the CULTURAL systems" (October 5, 2013, *Huffington Post*), arguing that racial disparities in incarceration were not caused by the War on Drugs, but rather "one parent families, third generation welfare families, hip hop music, ghetto unemployment and lack of economic investment in their communities, high crime in lower class communities, poor educations, lack of home discipline and love" (October 5, 2013, *Huffington Post*). Another explained racial inequality in the criminal justice system by pointing to a litany of racist tropes and stereotypes:

> The 70% illegitimacy rate in the black community, the lower high school and college graduation rates which happen to tie into being born into poverty to single mothers who then live on public benefits. These irresponsible, single teen mothers, living on welfare, food stamps, Medicaid, and Section 8 set a terrible example for these poor, innocent children. Children learn from their environment and these girls and deadbeat fathers don't set a proper moral example, they don't promote the importance of education as a means of escape from poverty, and they don't promote a strong work ethic because most of the single mothers and sperm donors don't work themselves.
> (October 4, 2013, *Huffington Post*)

Many simply argued that the actions and behaviors of blacks were the sole cause of the racial disparities in outcomes of the policies and practices of the War on Drugs. One commenter stated, "It's behavior. Period" (October 4, 2013, *Huffington Post*). While another argued, "It ain't the color of your skin, it's your actions that whites are bigoted and prejudiced against" (November 2014, *Huffington Post*). Along similar lines, commenters argued that the "bad values" of African Americans were a root cause of racial disparities in nonviolent drug arrests and incarceration. One commenter stated, "Young black boys and men need more self discipline," and claimed that within the black community, "there's way too much emphasis on fast money and a lifestyle that supports this" (October 5, 2013, *Huffington Post*).

Comment authors saw clear evidence of systematic oppression as instead serving as evidence of moral inferiority or social deviance. However, a case for why dominant groups are deviant or immoral could just as easily be constructed by selectively presenting trends or speculative observations (Kelley 1997; Hughey 2012a). For example, white college students are far more likely than their black peers to engage in binge drinking and dangerous activities while drunk such as driving, injuring themselves or others, fighting, unprotected sex, and "blacking out" (Seibert et al. 2003). These are harmful activities that many would regard as irresponsible, if not immoral. Yet, we do not have much of a dominant definition of white college students as dangerous or dysfunctional. In other words, the crucial difference in which narratives become dominant is which groups have the social power to make these labels stick.

Within comments in response to racial dynamics within the War on Drugs, racism or racial discrimination were routinely either seen as present but not consequential or completely nonexistent in contemporary society. Similarly, sociologist Eduardo Bonilla-Silva (2014) uncovered that whites often depict racial discrimination and differences in resources and opportunities as either an issue of the past or entirely eclipsed by other factors in determining people's outcomes and experiences. Many comment writers acknowledged that racism exists in some form, but still misrecognized the presumed dysfunction of people of color as the fundamental cause of racially disproportionate outcomes in the criminal justice system:

> Racism exists but it's not always the go-to reason, particularly in a society that for the past 50 years has gone above and beyond not only to attempt to equalize the races, but also to make concessions, excuses, special classes, and in some cases lower standards for minorities.
> (October 5, 2013, *Huffington Post*)

> Only a fool would say that blacks in general aren't born into challenging situations and that urban life is equal to suburban life... However, no one is FORCED to live a life of crime, it is a choice.
> (October 4, 2013, *Huffington Post*)

Individuals also employed a definition of racism as individual prejudice. This idea opened up the possibility of claims that racism against people of color was occurring alongside "reverse racism." These claims are part of a significant, well-documented trend in which, despite all contrary evidence, whites see themselves as either similarly wronged and victimized by discrimination to other groups or even more so (Norton and Sommers 2011; Hughey and Park 2013; King 2017). Additionally, commenters defined those who made claims about racial injustice experienced by people of color as "racist": "Shut up you racist fool. White people deal with this just as much as anybody else" (2014, *CNN*). Others saw the act of bringing up that topic as creating undue conflict: "Thank goodness for racial divisiveness articles. We wouldn't want folks to get along now would we" (October 28, 2014, *Huffington Post*). And commenters often downplayed the effect of discrimination and institutions on racially unequal outcomes through claims that whites also experience racism in the form of negative attitudes from people of color (see Doane 2006). After identifying himself as white, one commenter countered claims about racism as systematic oppression by stating, "I see racism every day when I go into a local Walmart and a black person looks at me like I just killed their dog after smiling at them" (October 29, 2014, *Huffington Post*).

While some commenters who rationalized racial disparities in the War on Drugs acknowledged that some form of racism exists in contemporary society, many others engaged in the complete denial of racial discrimination or bias, including reframing claims of racism as exaggerated or dishonest. Several comments implied the existence of an elaborate scheme among "liberals" and "black leaders" to make people of color "oversensitive" about racism. These claims framed understandings of racism as a fundamental cause of racial inequality as contradicting "facts" or "statistics" (which these commenters generally did not elaborate on):

> This is another attempt by the black grievance industry to capitalize on the fact that most people, especially liberal journalists, don't understand statistics, causality, and the whole post hoc ergo propter hoc fallacy.
> (March 9, 2014, *Huffington Post*)

> u all never like facts maybe if the first word out of your mouth wasn't racism all the time that could be a starting point but blacks have been brainwashed to yell racism to everything thank u al [Sharpton] thank u jj [Jessie Jackson].
> (October 4, 2013, *Huffington Post*)

Other claims hinged on assumptions about the intrinsic fairness of the criminal justice system:

> I don't think the cop said to his partner hey theirs [sic] a black guy put him in jail?
> (October 5, 2013, *Huffington Post*)

Cops and state attorneys have to prove cases, not simply say this guy is guilty because he is black... The antiquated white guilt of the boomers is really really old...
(April 8, 2013, *Huffington Post*)

Where do you get that? It's just a silly slogan that Black=guilty. I've seen no evidence of that during my time working for a judge.
(October 4, 2013, *Huffington Post*)

Racialized victim-blaming naturalizes the racially unequal outcomes of the War on Drugs. It depicts all those negatively impacted as criminal, deviant, and pathological; in short, as inherently immoral or inferior. Many of these claims also erased the influence of power dynamics, or group and individual-level differences in influence and resources, and the ways those differences are maintained, and social and historical context in understanding these issues. As we covered in the previous chapters of this book, the bulk of social scientific and historical evidence shows that racial double standards in the applications of drug laws and their enforcement shape these outcomes. The War on Drugs is part of the continued legacy of the criminal justice system as a means of social control targeting marginalized racial groups.

Colorblind Defense Frame

Commenters also came to the defense of the War on Drugs in ways that were not, on the surface at least, about racial meanings. I call this the **colorblind defense frame.** Audience members often wrote comments that pointed out the harms of drug use to rationalize the continuation of either a full-blown War on Drugs or at least some mild form of prohibition. One stated that drugs "are illegal for a reason, they are not good. They are a cancer on the people in this country. How can you put a price tag on doing the right thing?" (April 8, 2013, *Huffington Post*).

Pointing out that drug use poses a danger to individuals and society is not in itself a racialized claim. Indeed, most of these claims did not evoke racial categories. However, this theme did take on racial meanings in surprising ways. One commenter argued that legalization might make communities of color dysfunctional by increasing access to harmful substances: "[W]hen drugs that sedate are made legal to the poor and disadvantaged, it can lead to a lose [sic] of drive and initiative. I fear that stagnation in the Black community will be one of the many unintended consequences of legalization" (October 29, 2014, *Huffington Post*). Another commented that if drugs become legal:

> [...] there will be a bigger drain on the economy to support the abuse of all the druggies through the federal programs that are in place such as food stamps help with housing welfare medical assistance etc. It will put a heavier burden on the taxpayers in America that's exactly what Obama wants plus less able bodied people to fight a Corrupt Government taking from the hard working tax paying people and giving to the slacker's and Illegals entering through UNPROTECTED BORDERS.
> (October 21, 2014, *Huffington Post*)

While expressed with varying degrees of empathy and different political stances, both comments reflect anxieties over the possibility that the War on Drugs protects society from pathological or dysfunctional racial others.

Another commenter perpetuated the myth of "crack babies" toward this end. They stated, "For those who want to legalize everything – consider this: Who is going to take care of all the 'crack babies?'" (April 8, 2013, *Huffington Post*). Fears about so-called "crack babies" reflect a racialized moral panic about the neonatal effects of crack-cocaine use. In a review of the available medical evidence, Deborah A. Frank, a children's health researcher, and colleagues (2001:1613) reported "no consistent negative association between parental cocaine exposure and physical growth, developmental test scores, or receptive or expressive language." Regardless, the "crack baby" myth spread like wildfire. It aligned with negative stereotypes about black women as dysfunctional mothers. It enabled depictions of children as the innocent victims of drug addiction in need of saving connected to the broader racialized moral panic of the "crack epidemic" (Sandy 2003). It, therefore, proved successful in helping rationalize targeting black communities for invasive and violent policing and mass incarceration under the banner of drug abuse prevention (Sandy 2003).

Commenters also routinely made claims about free will and choices that did not evoke racial categories but still had the effect of rationalizing the outcomes of the War on Drugs as legitimate reflections of people's decisions to engage in drug use or crime. This theme is somewhat similar to what Eduardo Bonilla-Silva (2014) calls "abstract liberalism" or using arguments about individual freedom and responsibility and the assumption of equal access to opportunities to explain the existence of racial inequality. Interestingly, this approach also constitutes a form of victim-blaming but without mentioning race.

One commenter drew upon the notion of personal responsibility to argue, "Nobody is ever held accountable for their bad behavior in liberal fantasy land. In the real world, we have a saying.... If you can't do the time, don't do the crime" (April 8, 2013, *Huffington Post*). The idea of personal responsibility helps people rationalize the racially unequal outcomes of the War on Drugs, such as racialized mass incarceration because it credits them to the individual decisions of blacks (see Emerson 2011). Comments in this theme reflect a lack of sociological imagination or ability to connect individual circumstances to social and historical trends (Mills 1959) by emphasizing freewill but ignoring the broader social context within which people exercise it. As one commenter argued, "Nobody is pointing a gun at you and asking you to smoke pot" (April 8, 2013, *Huffington Post*). Another contended, "It doesn't matter how you FEEL or that you like drugs, they broke the law and have been placed where they belong. They should never get out" (April 8, 2013, *Huffington Post*).

These claims were often ostensibly nonracial. But commenters deployed them in response to data and arguments emphasizing the racial unfairness of the policies and practices of the War on Drugs. These claims implied that racial disparities in incarceration rates accurately reflect the aggregation of individual decisions to commit crimes rather than differences in institutional treatment or access to resources: "Here's a thought....these people wouldn't be going to prison if they weren't committing crimes. Stop that first and then watch the other statistics fall away" (October 5, 2013, *Huffington Post*).

Another commenter wrote, "if people get stopped and frisked or pulled over, give them no reason to arrest you. If people obey the law, they won't end up in prison" (October 5, 2013, *Huffington Post*). Comments like this parroted the "do the crime do the time" discourse, yet flouted relevant trends like the extreme racial disparities in the likelihood of being pulled over or stopped and frisked by police and the enactment of racial profiling through these policies

(Welch 2007; Epp, Maynard-Moody, and Haider-Markel 2014; NYCLU 2016).

In its most extreme form, the narrative of individualism and personal responsibility even manifested in overt and intense calls for toil and punishment, torture, and even murder of those placed in the criminal justice system. One commenter expressed a desire to not only double down on the current policies and practices but exclaimed:

> We need to make jail harder. Get rid of A.C., cable and all that nice crap that the rest of us have to work for. You want flavor in your food? Go get a job and buy your own flavor! Teach them that jail isn't a good place where they can go spend some time. Work their asses off. I'm talking 12–16 hour days. All this crap about 'inmates don't have rights....'. You're damn right they don't! They took someone else's rights and so now they forfeit their own. You want 'em back? You'll get 'em back when the system sees that you know how to use and respect them. Right now, you go dig a ditch!
> (April 8, 2013, *Huffington Post*)

Comment authors also depicted jail and prison inmates as lazy free-riders: "Time to give our inmates hard labor to make them pay their own way in this world, and not the taxpayer. No work, no eat. It's that simple" (October 19, 2013, *Fox News*). And some commenters even called for the murder of drug dealers and users. One argued, "Selling drugs to our children is the same as ASSAULT if you ask me ... (physical harm). Why not just EXECUTE all incarcerated murderers immediately. This would open up space for those that assault our children" (April 8, 2013, *Huffington Post*). Another stated, "kill em.... kill every drug addict stupid enough to do the stuff" (April 8, 2013, *Huffington Post*). In short, ideas about racial groups, morality, victimhood, and responsibility played a significant role in resistance to critiques of the War on Drugs within the debate.

Conclusions

We now have a much more complete sense of how people construct and express racial and political identities through their participation in the War on Drugs debate. The unique frames that only appeared in comment sections and not newspapers suggest that the effects of framing and agenda-setting from mass media on how people talk and think about contested social issues are not unidirectional. Ongoing discussions through which people distinguish and define themselves and others also shaped the perspectives that people have on these issues (Walsh 2004). Regardless, media discourse about contested social issues is significant. The War on Drugs debate in mass media communicates to audiences what matters about this issue and why.

In Chapter 3, I argued that the frames and themes in the newspapers analyzed tended to imply that the racially unequal outcomes produced by the War on Drugs were either legitimate and inevitable or not important. So, what role does racial identity and ideology play here? It's not hard to imagine that these messages still hold resonance with those who used a racialized victim-blaming frame to understand these outcomes in online comments. Mass media coverage of the debate provides ample content that audiences can talk back to in various ways. And the act of talking back to the variety of framed messages in the media allows people to assert their identities and reinforce their conception of the world.

Overt racial claims tended to hold less resonance in the debate. And racial silence, or the omission of racial and particularly racial justice claims, allowed claims to resonate with audiences who hold attitudes of **racial apathy** (Forman and Lewis 2006), and therefore don't care about racial inequality or, albeit less commonly, overt bias or hatred toward people of color. We can see with much greater depth and clarity how many white people respond when others break the norm of racial silence and critique the War on Drugs as a matter of racial justice. While they varied in whether they were overt or covert about the use of racial categories, these responses employed many ideas akin to colorblind ideology, especially abstract liberalism, cultural racism, and minimization of racism (Bonilla-Silva 2014).

In the debate over the War on Drugs in online comments, people routinely explained evidence of racial oppression in ways that depicted these outcomes as natural or legitimate. Many commenters countered evidence of institutional racism by reframing it as reflecting the pathology, criminality, and inferiority of impacted groups and people within an otherwise just and fair system. These perspectives have emerged as dominant forms of racial discourse now that racist claims about biological or genetic group differences have lost influence (Steinberg 1989).

Oscar Lewis, an anthropologist, in his 1966 *Scientific American* article "The Culture of Poverty," argued that the persistence of a lower-class status could sometimes result from an ethnic group developing what he called the **culture of poverty**. This term refers to a pathological subculture that prevents upward mobility even when structural changes produce new opportunities. He based this assumption on cursory observations of ethnic groups in various countries that he felt exhibited hopelessness, failed to live up to middle-class values, had mostly female-headed families, and possessed a lack of impulse control. Similar ideas were often used within the debate over the War on Drugs to depict the disparate criminalization of black and Latino people as a product of their assumed dysfunctional cultural practices or values.

So is inequality maintained by the "bad culture" of marginalized groups, as many commenters claimed? Let's look at the flip side of this claim: 96% of the top 1% of income earners are white, and whites own 90% of the total wealth in the United States (Brantley 2011; Bruenig 2014). In a series of recent social psychology studies, the wealthy expressed lower levels of generosity, trust, and empathy, and were more likely to steal, lie, cheat, approve of unethical behaviors in the workplace, engage in victim-blaming, and exhibit narcissistic tendencies and a sense of entitlement (e.g., Kraus, Côté, and Keltner 2010; Piff et al. 2010, 2012; Piff 2014; Kraus and Callaghan 2016). Yet, dominant group members downplay the existence of these antisocial and harmful attitudes and behaviors. In contrast, the idea that disadvantaged social groups are culturally or morally inferior helps rationalize their ongoing subjugation (Bonilla-Silva 2014).

Many claims in the debate that employed racialized victim-blaming explicitly focused on ideas about not just culture but racial differences in patterns of family formation. These claims echo a 1965 US Department of Labor report titled "The Negro Family: The Case for National Action" by sociologist, Assistant Secretary of Labor, and eventual New York State Senator Daniel Patrick Moynihan. The report presents some statistical data on the characteristics of black families in the United States. Then, drawing on speculation about the importance of father figures influenced by the ideas of Sigmund Freud (Greenbaum 2015), it proposed that the structure of what he called "the Negro family" was inherently dysfunctional. From there, Moynihan (1965:47) infamously argued that "the present tangle of pathology is capable of perpetuating itself without assistance from the white world." As Stephen Steinberg (2011) points out, "Moynihan made the fatal error of inverting cause and effect." In this tangled interpretation, patterns in family structure have a life of their own, independent from historical, economic, social, or political contexts.

Nuclear, male-headed families were never a product of superior culture. These types of families are, in fact, historically and geographically novel. They were made possible by increased opportunities for white couples to purchase their own homes and other economic, social, and political resources that came with their increased class status and the construction of predominantly white suburbs in the postwar boom of the 1950s (Coontz 1992; Greenbaum 2015). In contrast, extended family networks, flexible household arrangements, and tightknit community relationships have provided highly functional strategies for making the most of the diminished access to resources in many black communities (Stack 1974).

Like many of the comment writers, Moynihan's proposed solutions avoid any real reckoning with the ongoing impacts of centuries of well-documented racial discrimination. Both Lewis and Moynihan take a tone of being concerned with addressing social problems such as poverty or racial inequality, just as many commenters were ostensibly worried about issues like crime or drug use. However, the cultural pathology arguments advanced by both influential works share much in common with the claims that still permeate our debates about racialized and contested social issues. In consequence, they rationalize and naturalize racial inequality and stifle racial justice-oriented social reforms.

Along similar lines, other commenters employed concepts like personal responsibility to shift blame away from an acknowledgment of poverty, racism, state violence, public policies, and economic arrangements to the individual. The remaining simply denied or minimized the impact of racial discrimination in producing these outcomes. Participants also derided the breaching of racial silence and the evocation of claims about the racial injustice of the War on Drugs as divisive, "reverse racism," or unfairly scapegoating innocent whites.

Code words such as "crack babies," "welfare recipients," and "little street corner punks" frequently appeared in online comments critical of the War on Drug. Alongside equating black and Latinx people with dysfunction or pathology, racialized imagery of threat via references to "terrorists," "cartels," "thugs," and "street gangs" was frequent in both comments and newspapers. These code words allowed the proponents of both sides of the debate to associate the War on Drugs to the protection and continuation of the white-dominated racial order from "outsiders". They also suggest that black and Latinx people were unworthy of empathy and social investment or inherently menacing. As we discussed in the last chapter, these tactics reflect the ways that implicitly racial discourses from political elites (see Mendelberg 2001; Haney López 2014; Hughey and Parks 2014) influence the thought and speech of everyday people.

Code words enable claim makers to construct racialized subject-positions while maintaining surface-level racial silence. The debate over the War on Drugs within print and digital media presents a site of identity construction. Ideas and narratives that people can use "to define someone else, to make 'their' identity in the shadow of 'ours'" (Matheson 2005:142) and demarcate legitimate claims serve as essential tools for racial identity construction. In both newspapers and online comments, statements about the legitimacy of other claim makers such as "liberals," "conservatives," "blacks," "whites" helped individuals construct in-/out-group distinctions and position their sense of self in relation. Racial and political categories were crucial in this process. The War on Drugs debate is a space for deeply political contestations over the definition of racial groups such as white or black and what it means to inhabit them.[14]

Writing about television news content, Robert M. Entman and Andrew Rojecki (2001:57) cautioned that when audiences consume media content that centers whites' experiences as the norm, contains implicit references to racial stereotypes, and omits vital context or discussion about racial justice, it "may work against the development of greater interracial empathy and trust." Within interactive digital media, identity construction in relation to mass media content takes place not through the consumption of media content but in how people formulate responses and personal positions on the narratives they encounter. By implying people of color rather than racial oppression as the fundamental problem of the War on Drugs, media content containing racial silence and code words helps whites to draw racialized symbolic boundaries. These boundaries create barriers to empathy to people of color impacted by these policies and practices in the face of evidence that links the War on Drugs to structural racism.

In the debate on the War on Drugs, people engaged in racial identity construction through moral and racial boundaries. They defined racial differences as meaningful reflections of moral differences and by creating a sense of group position. Within comments, whiteness was very often defined by and contrasted with blackness. Commenters then connected this black/white binary to a binary of moral values such as innocence/criminality. In response to evidence of grave racial injustice in the application of drug laws and criminal justice practices, audience members presented white racial identities as morally superior. They interpreted racialized criminalization as a natural and legitimate outcome of the assumed pathology and moral failings of black individuals rather than systemic racism. Racial justice critiques of the War on Drugs did not resonate with these presumably white audience members. A pattern of racial silence and code words was so prevalent in the media content analyzed because critical racial justice claims contradict the racial meanings and identities of audience members such as those who define whiteness as inherently moral, functional, and law-abiding.

Discourses don't have to be dominant to allow individuals to assert identities in ways that support their worldview. They can be quite critical of the status quo. As we saw in many of the claims in the racial unfairness frame in online comments, interactive media does provide space for the construction of racial identities through subject-positions that center or empathetically embrace people of color. These types of claims present **counterstories** or narratives and forms of storytelling about events or issues that reflect and respect the lived experiences, knowledge, and worldviews of oppressed groups such as people of color (Deglado 2000; Solórzano and Yosso 2002). As legal scholar Richard Delgado (2000:60) argues, while dominant groups often tell stories that affirm their identities and naturalize their social positions, "the stories of outgroups attempt to subvert that reality."

These claims can also be considered **counter-framing**. They contradict the frames that other readers employed to rationalize racial inequality. As noted by Joe R. Feagin (2010), a sociologist, there is an enduring American tradition of counter-framing by people of color in the United States as part of the struggle for racial equality. It enables them to identify racial oppression and the practices that maintain it as a social problem and cause of inequality. For example, many commenters defined blackness as innocent or unfairly victimized and whiteness as either ignorant of or unconcerned with structural racism by referencing personal stories and experiences, historical and contemporary trends, or even just with the language and definitions that they employed. However, such claims often lack resonance with those who hold dominant racial meanings. This lack of resonance presents a severe challenge for individuals who want to bring critical perspectives about racial injustice to light.

Bringing all this together, the production of media content on the War on Drugs and identity construction among audiences contribute to a cycle that helps white Americans rationalize or ignore the causes and consequences of institutional and structural racism. These dominant sets of ideas about race and racial inequality don't just allow people to justify racially unjust arrangements like the War on Drugs, they also provide discourses that people can employ to refine and assert their identities and communicate a sense of who they are. Perhaps this is why debates about significant social issues often feel so personal; claims in these contestations can reaffirm or challenge people's deeply held sense of self and social position.

Discussion Questions

1 Identify a contested social issue that generates strong feelings in you. Explain. How might your identity affect how you perceive this issue?
2 Identify a social problem about which you care. Now, how can you make compelling arguments about this social problem that would appeal to people who have no personal interest in this problem?
3 What offline and online spaces do you use to engage in discussions about current events and other issues? Do you discuss racial issues in these spaces? If so, how often? Why or why not? Explain.
4 To what extent are the spaces where you discuss issues segregated by age, race, class, political orientation, or other social characteristics? Explain.
5 When and how often do you read or watch news media? On what platform (cell phone, Facebook, Twitter, Gmail, blog, etc.)? What topics do you typically focus on when selecting news? Do you consider the source? Do you consider the narrative presented and whether it fits in with your worldview?

6   Some news outlets are reconsidering whether to retain the comment fields accompanying their stories. For particular content deemed controversial, it is more likely that comments will be closer or heavily moderated. What purpose do you think the comment sections on online news sources provide?

Notes

1 Erving Goffman's (1959) dramaturgical theory of social life suggests that social interaction operates similar to a play on a stage, including scripts that we follow, supporting casts and props that help make our performances more convincing, and, most relevantly, frontstage (formal interactions in the presence of strangers and public audiences) and backstage (private informal settings). Sociologists have demonstrated how conversations that take place backstage between dominant group members about racialized and gendered issues play an important role in how they form identities and that these interactions often reinforce or rationalize social inequalities (e.g., Meyers 2005; Picca and Feagin 2007; Hughey 2011; Embrick and Henricks 2013).
2 See Anderson (2016) for more on the history and present of white backlash to black equality and progress.
3 This idea derives from the philosopher Louis Althusser ([1971] 2001:117) who wrote that "all ideology hails or interpellates concrete individuals as concrete subjects, by the functioning of the category of the subject." This statement may make little sense at first glance but it suggests something important. Althusser means "hail" in the literal sense and asks us to imagine the following scene: you are walking outside and hear a voice call out to you (maybe something like "hey you!") from behind and so you twist your head or perhaps even your whole body in recognition (Althusser [1971]2001).
4 George Herbert Mead (1934:201) writes: "it is the characteristic of the self as an object to itself that I want to bring out. This characteristic is represented in the word 'self,' which is a reflexive, and indicates that which can be both subject and object."
5 For example, Erving Goffman (1967:111) noted that often "one builds one's identity out of claims."
6 In some ways, this is an old idea, dating as far back as Emile Durkheim's ([1912] 1995) insinuation that in the early 20th century, the categories that people use in thinking about the world around them, such as distinctions between "sacred" and "profane" in religious life, are really the products of society.
7 For more background on the concepts of "discourse(s)" and "subject-positions," see Foucault (1971), Weedon (1989), van Dijk (2008), and Hall (1995, 1996, 1997).
8 For instance, in the early 1900s, sociologist Charles Horton Cooley (1906) argued that the sense that you are distinct from the outside world around you is not something innate to humans but something that people learn and develop over time.

9 W.E.B. Du Bois was one of the earliest and most influential scholars to refute biological and genetic views of race. He demonstrated that harmful effects of "scientific racism," which was being taught in all of the elite schools and rationalized racial oppression (Taylor 1981). Disturbingly, there has been a resurgence of scientific racism in the age of biogenetics and its pernicious effects have been analyzed (Roberts 2011; Phelan, Link, and Feldman 2013; Byrd and Hughey 2015; Byrd and Ray 2015).
10 The technique I used here is theoretically informed or purposive sampling (see Luker 2008:104; Altheide and Schneider 2013:25, 55).
11 Collecting online comments was somewhat hampered by the ways that news websites "deal with offensive comments, including turning 'comments off,' not archiving comments, and adopting aggressive comment moderation policies" (Hughey and Daniels 2013:332). Regardless, 24 articles were found using the criteria of relevance and including only those from US sources that contained variously racialized critiques of the War on Drugs and possessed over 20 comments.
12 While other comments include the full date of when they were posted, the comment section on Forbes.com does not catalog this information in full.
13 As noted in the first chapter of this book, the establishment of marijuana prohibition and racism against black and Latino people in the early to mid-20th century was deeply intertwined (see Provine 2007; Alexander 2012).
14 See Hochschild (1999) on "cultural warfare."

Chapter 6

Conclusion

The debate over the War on Drugs stands at the crossroads of a multitude of contentious, impactful, and problematic issues in American society, including racial oppression, mass media and digital communication, and political conflict. With that in mind, the goal of this book has been to impart a sociological perspective on the ways that social, cultural, and political dynamics crosscut the debate itself through empirical data and in-depth analysis. Along the way, we have explored several topics. We have looked at how issues become identified as social problems. We have examined how the War on Drugs was declared and developed into a set of policies and practices. We have analyzed how the media influences public agendas and debates. And finally, we investigated how people form identities in taking positions on contested social issues. Throughout it all, we focused on how each of these aspects of the debate reflects critical issues of power, meaning, and racial inequality in our society.

I hope that you now know more than you did when you first picked up this book and that this newfound knowledge has nourished your ability to envision how historical and social forces impact people's everyday lives. But knowledge, while important, is not enough. The information contained in these pages can also enable us to become more sociologically mindful (see Schwalbe 2005) in the actions we take in our everyday lives. And so, in concluding this text, I want to shed light on how the findings and contributions of this book provide ways for reframing the debate, discounting racial meanings that justify inequality, and working toward structural changes that reflect a profoundly sociological awareness of these issues.

Revealing Racial Meanings from Ideology to Identity

If you read through the significant research on race and racism in the field of sociology, you will likely come across two concepts repeatedly: ideology and identity. These concepts not only both begin with the letter I, but they also both help explain the role of cultural and interpretive processes of meaning-making in the reproduction of racial oppression. In short, they describe forms of **racial meaning**. They are interconnected aspects of how race and racism manifest as a system of cultural representation. Racial meanings allow people to interpret or make sense of themselves and the world around them in racial terms.

We discussed both racial ideologies and racial identities in the previous chapters of this book, but we don't usually see their connection. Sociologists often explain the role of meaning in the reproduction of racial inequality by only referencing one or the other. However, we need to think of them as just different types of racial meaning. Throughout this book, I have brought to light the patterns in racial meanings within arguments for and against punitive drug policies and explanations of racial disparities in its application. Examining the War on Drugs debate in print and digital media helps us understand how people develop and utilize dominant racial meanings in contestations over racialized social issues. **Dominant racial meanings** represent the dominant racial group's perspective and provide ways of justifying that social position within the system of racial inequality.

Ideologies serve a specific function. They allow people to rationalize the present social structures even if they live in societies wrought with inequality, conflict, and oppression. The dominant racial ideology is the specific set of ideas that are most commonly used in a particular era to justify the existence of material inequality between racial groups, racial oppression, and the practices that maintain it (Bonilla-Silva 1997). Dominant ideologies change over time. They reflect the social arrangements of a given society in a historical or contemporary period. As we discussed in Chapter 2, colorblind ideology (see Bonilla-Silva 2014) employs modern ideas such as equality of opportunity, individualism, and avoidance of overtly racist language. It represents the dominant racial ideology.

Dominant ideologies offer people general sets of ideas and narratives. But, in the complexity of everyday social interactions, they are communicated and employed in context-sensitive ways. Amanda Lewis (2004:633) argues that racial ideologies "provide ways of understanding the world that make sense of racial gaps in earnings, wealth, and health such that whites do not see any connection between their gain and others' loss." Similarly, they enable rationalizing interpretations of the racially disparate practices, policies, and outcomes of the War on Drugs.

In the debate, people used ideas and narratives to present racial disparities in the outcomes of the War on Drugs as rational or just. Participants in the discussions often used dominant racial ideologies for victim-blaming in response to evidence or claims of racial oppression. Three frames of colorblind ideology (see Bonilla-Silva 2014) influenced how people rationalized the racial injustice of the War on Drugs. Cultural racism, which points to the assumed cultural practices or values of a group to explain their social position, was often explicitly expressed in ideas connecting black or Latino culture to criminality or pathological family formations. Many debate participants' arguments explained these outcomes as a matter of personal responsibility employing abstract liberalism or the idea that social positions reflect individual effort and choices in the context of freedom and equal opportunity. And the minimization of racism was explicitly seen in people's claims that the criminal justice system is fair and objective.

However, people did not just make claims about the War on Drugs but also themselves, others, and US society. Dominant racial ideologies not only have the function of rationalizing specific outcomes or aspects of racialized social issues, but they also "enable people to understand and to accept their positions within a stratified society" (Lewis 2004:623) through the process of identity construction. Racial ideologies contain racialized subject-positions (Hall 1997) or "vantage points" (see Davies and Harré 1990) from which to view a racialized social issue. When people adopt such subject-positions, they connect their stance on that issue to particular understandings of the meaning of racial categories and the characteristics of the people occupying those categories. These interpretations, understandings, or perspectives enable them to form a sense of who they are concerning others and the world.

Cultural hegemony, a concept developed by the early 20th-century Italian political theorist Antonio Gramsci ([1929–1935] 1971), helps us understand the relationship between social group power and cultural values and norms in unequal societies. Gramsci pointed out that dominant groups hold influence over mass culture in ways that help normalize their authority and idealize their interests and perspective (Lears 1985). Rather than being overtly coerced through violent force, people often willingly consent to power arrangements because the dominant cultural context makes them seem inevitable or natural (Lears 1985). Even if individuals personally disagree, cultural hegemony produces a kind of shared common sense about which activities, ideas, social categories, or social arrangements are normal, real, or ideal.

This relationship between power and culture also influences how people define themselves or construct identities. Power dynamics shape racial categories and their commonly held meanings in society. Just like colorblind ideology is the dominant ideology, hegemonic whiteness is the dominant racial identity. Per this hegemonic or dominant racial meaning, what does it mean to be white? **Hegemonic whiteness** describes a racial identity rooted in the assumption that being white is a culturally normative state and a naturally dominant social position (Lewis 2004).

Hegemonic whiteness is part of the symbolic dimension of racial oppression (see Collins 1993). It describes the way that the commonly held ideas or associated characteristics of whiteness influence people's lives. In other words, it is the way that people perform or act out the dominant image of what it means to be white. Not all stereotypes or commonly held images of a racial group are negative or denigrating (Henricks and Embrick 2013). In the dominant image of whiteness, being white is idealized as a state of normalcy, morality, innocence, purity, victimhood, or authority. While these are varying and even seemingly contradictory definitions, they all help position whiteness as an idealized or superior state of being in cultural expressions and media depictions. In everyday life, these images of whiteness are put into practice in ways that maintain systems of racialized social control such as the War on Drugs.

When people construct racial identities through hegemonic whiteness, they draw out what it means to be white in comparison to the assumed characteristics and images of both racialized minorities and whites who fail to live up to this ideal (Hughey 2010). Consider this quote from the last chapter:

> Criminals tend not to understand the concept of delayed gratification and look for easy money. I get a kick out of people who make fun of white people because we're boring or

stiff and take school seriously, go to college and take that seriously, have two parents raising kids, obey the law, know the law, etc.

In one sense, this statement reflects the dominant racial ideology. It presents the racial inequalities and consequences produced by the War on Drugs as a natural or justified outcome of assumed nonwhite criminality. In another sense, it also provides an opportunity for identity construction. The comment defines whiteness and therefore the commenter's sense of self as connected to innocence and moral superiority and nonwhiteness as criminality and moral inferiority. In sociological jargon, the discourse used by this commenter produced a racialized subject-position. They then used this subject-position to identify themselves. Racialized subject-positions connect the function of ideology to the process of identity construction.

Many white people are reliant on the idea that the racially unequal outcomes of social arrangements such as the War on Drugs are legitimate reflections of racial inferiority and superiority for their very sense of what it means to be white. These dominant racial meanings contain distorted forms of reasoning like what social scientists and philosophers call the **just-world fallacy** or the assumption that the bad things people experience, such as violence, stigma, criminalization, poverty, or oppression, are inherently deserved (see Lerner 1980). When people explain how the world works by imagining that bad things happen to bad people and good things happen to good people, they avoid feeling discomfort about the unfortunate, arbitrary, and tragic situations of others, but they often misunderstand social reality (Lerner 1980).

Dominant racial meanings allow people to develop a sense of self. They provide many people with what Anthony Giddens (1991), a sociologist, calls **ontological security** or a sense of continuity and comfort in the flow and meaning of events in one's life. In philosophy, **ontology** is concerned with questions of being. In this case, ontology includes how people answer questions about what it means to *be* white, black, or of any other racial category. When people answer this question, they make an **ontological investment**.

For many whites as well as people in general who buy into dominant racial meanings, evidence of racial oppression feels like an attack on their personal sense of identity. As we've seen throughout this book, research overwhelmingly demonstrates that the War on

Drugs is part of a system of racialized social control. However, this information contradicts and disrupts the ways that many people, and especially many powerful people, make sense of themselves, their everyday activities, and their social position.

Mass media has been a major site for the debate over the War on Drugs and a host of other racialized and contested social issues. As we can see in Chapters 3 and 5, processes of media production and audience reception and interpretation are shaped by dominant racial meanings. Social institutions such as mass media provide pathways through which dominant racial meanings can travel and spread. Hence, public debates over racialized and contested social issues such as the War on Drugs are often more like heated exchanges (in the case of online comments) or constrained and muted dialogues (in the case of newspapers) than discussions that light a path to appreciating and addressing their causes and consequences.

Given all this, we may feel helpless. How can we engage with and transform the reality that systems of racial oppression, which manifest in policies and practices like the War on Drugs, create arbitrary suffering and unjust bondage and limit people's freedom, opportunities, and life chances?

Challenging Commonsense Myths

> A long and darkly shrouded history of hierarchical power arrangements has given shape to what we today commonsensically consider normal and what we control against as deviant. By demythologizing and thereby destructuring the oppressive bondage of this common sense, we are able to partially recover our freedom of thought and action.
>
> Stephen J. Pfohl (1985:383)

In Chapter 1, I demonstrated how the policies and practices of the War on Drugs have their genesis in moral panics – the labeling of a group or activity as threatening. More specifically, they are rooted in racialized moral panics. Powerful actors in the realms of politics, mass media, and government equated a racial group with some substance or drug-related activity and presented this combination as worthy of alarm and fright to white citizens. This background gives us some important insights for understanding the types of logic that influence our contemporary discussions about the War on Drugs.

When we peel back the surface of racialized moral panics over things like drugs and crime, we find **myths** about social groups and society. Philosopher Roland Barthes (1957:143) wrote:

> myth does not deny things, on the contrary, its function is to talk about them; simply, it purifies them, it makes them innocent, it gives them a natural and eternal justification, it gives them a clarity which is not that of an explanation but that of a statement of fact.

In evoking the idea of myth, Barthes isn't referring to ancient origin stories starring mystical beings, but rather how language and power can present certain narratives and ideas as natural or a given. That is to say that myths operate through appeals to common sense. They remove all history and context and make things appear as though they are simply natural. When we encounter myths, they are presented as self-evident facts rather than coherent explanations for how the world around us works. Yet, as we saw throughout this book, myths do play a large role in how people explain things within debates.

Few people would argue about whether there is a racial disparity in who is punished by drug law enforcement in the United States. However, misconceptions and arguments abound about the causes and nature of this trend. Now having looked at the broader context and evidence, we see that this disparity reflects moral panics about race, drugs, and crime, racial double standards in the use of institutions of social control by predominantly white elites, and the unresolved consequences of racial oppression in the United States. This context, history, and evidence are documented in Chapters 1 and 2 of this book. But the contemporary debate over this issue is a whole other matter.

In Chapters 3, 4, and 5, we saw that within the debate in newspapers and online news comments, there was a consistent narrative and set of assumptions. Commentators commonly assumed that the racially unequal outcomes of the War on Drugs reflect some innate dysfunction and criminality of black and Latinx people and the innate moral and social superiority of whites was often accepted or advanced, even by those arguing against the War on Drugs. The acceptance of this narrative was stated in implicit and explicit ways. In other instances, silence about the racial justice implications of the War on Drugs or the causes of racially disproportionate outcomes did not fully challenge the legitimacy of this narrative.

This assumption about the causes and consequences of racial disparities provided a potent resource for predominantly white audiences of the media content produced by this debate to maintain their sense of racial identity. Racialized victim-blaming, othering, and the use of racial code words remain potent strategies to gain influence by drawing on commonly held myths about race, crime, and drugs (see Haney López 2014). By considering the power of these strategies, we can better understand why even arguments against the War on Drugs often perpetuate such myths or simply leave them unchallenged.

How do myths about race, drugs, and crime influence the public debate? Newspapers remained generally silent about relevant racial matters. Internet commenters on news stories espoused innocence, morality, dysfunction, and criminality as traits inherently associated with racial groups. And participants in the debate on both sides used code words to conjure racial imagery, yet avoid the direct evocation of racial categories. Looking at the big picture of the whole debate can help us see patterns. Knowing these patterns can make it easier to decode coded language or understand what is communicated through silences. All of this helps us engage in more conscious and informed social and political action. Because if we pay close attention, we can start to see these dynamics in how powerful individuals talk about these issues as much as non-elites that might be discussing the War on Drugs online or at their kitchen table. For instance, in 2016, Maine Governor Paul LePage stated the following in a town hall on the state's drug problem:

> These are guys by the name D-Money, Smoothy, Shifty. These type of guys that come from Connecticut and New York. They come up here and sell heroin, then they go back home. Incidentally, half the time they impregnate a young, white girl before they leave, which is a real sad thing because then we've got another issue we have to deal with down the road.
> (Phelps 2016)

The use of code words and the association of criminality and innocence with white and black to the point of having the terms stand-in for one another was something we saw in how commenters rationalized the racial inequality perpetuated by the

War on Drugs within the debate. The use of racialized nicknames and the juxtaposition of these supposed men to "young, white girl[s]" suggests that the "type of guys" discussed by LePage are not themselves white without evoking the racial category of "black."

And the response to criticisms of this statement further demonstrates how code words and silences enable deniability about breaking the vow of racial silence. A representative for LePage told ABC News: "Race is irrelevant [...]. What is relevant is the cost to state taxpayers for welfare and the emotional costs for these kids who are born as a result of involvement with drug traffickers" (Phelps 2016). For many of us, these comments are deeply offensive if not simply insensitive. However, these types of claims are also regarded by others as mere common sense,[1] especially considering their similarity to many claims in the public debate on the War on Drugs.

These types of ideas that pervade the debate on the War on Drugs are **commonsense myths** – ideas and narratives that don't reflect reality but become taken for granted and commonly accepted. These myths play a decisive role in not only political rhetoric and policies but also how everyday people think about racialized social issues. Consider how common the tropes of dysfunctional or dangerous racial others (i.e., "thugs," "inner-city gangs," "welfare recipients," "terrorists," "illegals," "crack babies," "cartels") were throughout the debate even among those who wanted to end the War on Drugs. Because these ideas and images so consistently and subtly appear in public discourse, statements that imply black or Latinx people as threatening or inferior unfortunately resonate with many white Americans as simply a direct articulation of what already seemed to be a matter of common sense.

As noted in Chapter 3, ideas, explanations, and stories resonate with people based on their traditions and cultural ideals. And these traditions and ideals are often connected to group interests. The commonsense myths found in the debate over the War on Drugs idealize the interests of white elites. In doing so, they help maintain social arrangements that advantage this group at the cost of the well-being of others.[2] Many of us want to believe that we live in a fundamentally fair society. This is a comforting notion but it confuses our ideals for reality. The history and ongoing reality of the War on Drugs and the public debate that surrounds it shows that these beliefs rely on narratives of victim-blaming. To see such overwhelmingly unequal outcomes as rational or just, people interpret the world in ways that shape how they see themselves and others.

We must take up the important work of W.E.B. Du Bois as described in Chapter 2 – challenging commonsense myths about the causes of racially unequal criminal justice outcomes whenever and wherever we encounter them in the media, spoken by politicians and other leaders, and even uttered by our friends and families. Building a more complete and mindful understanding of how racial oppression and racial categories influence people's everyday lives or cultivating "racial literacy" (Twine 2004) is essential for this task.

The power of commonsense myths is that they appear to be self-evident. So, in our everyday lives, we must engage in what sociologist Celine-Marie Pascale (2007:49) called "confronting what appears to be obvious" and "learning to see that which commonsense actively works to conceal," so that we can challenge the myths that support the racially unequal application of laws and distributions of opportunities, rights, and resources. Most importantly, we need to challenge racial silence that leaves these myths untouched. As the Civil Rights Movement leader Martin Luther King (1963) wrote from a Birmingham jail cell, "we will have to repent in this generation not merely for the vitriolic words and actions of the bad people, but for the appalling silence of the good people." Only when we challenge racial silence can we confront the various ways that the practices and policies of the War on Drugs comprise a system of racialized social control and reproduce racial inequality. But to truly address the root causes of these injustices, many of us need to shift how we think and feel and even our sense of who we are.

Rethinking Empathy, Morality, and Identity

It's not enough to simply be aware and share with others the historical and social context and the evidence that the War on Drugs is racially unjust or similar information regarding other racialized and contested social issues. The findings from online comments in Chapter 5 were littered with examples of how compelling evidence of racial oppression and injustice was contested, if not outright rejected, by others participating in the debate. So, let's return to a central issue that this book grapples with – empathy – to understand the debate over the War on Drugs.

On both sides of the issue, claims about the War on Drugs that defined black and Latinx people as a threat or problem seemed to overshadow those that defined them as the sufferers of injustice or oppression. Racialized narratives, categories, and images spread throughout the media and public debate over the War on Drugs enable individuals to avoid feeling the emotional weight of the suffering caused to those targeted by its policies and practices. Because this is a racialized issue, the debate and its outcomes reflect a **racial empathy gap**.[3] Empathy is the capacity to imagine yourself in the position or situation of someone else. Empathy allows us to think about how a situation or position might affect someone else or make them feel. A racial empathy gap occurs when this process is disrupted by dominant racial meanings and it can have intense, even life or death, consequences.

Consider the role that empathy plays in court cases. The capacity to empathize influences who juries and judges identify as capable of experiencing suffering, and so the racial empathy gap creates biases in sentencing and even in the use of capital punishment (Taslitz 2013). Given that whites are more likely to be in positions of decision-making in courtrooms, empathy tends to be extended toward whites rather than black or Latinx people in these circumstances (Taslitz 2013). Imagine that you are awaiting news that will affect the conditions for the rest of your life, maybe even whether you live or die. Now consider that in this situation, your fate is in the hands of a group of people who are not just motivated by the facts of the case but also whether or not they recognize you as someone who's suffering matters. And because of the images, stories, and ideas associated with the social group you belong to, some of those people are simply unable to do so.

The racial empathy gap also means that whites are less likely to recognize the capacity of black people to experience physical pain. In a recent experiment, white medical practitioners, when instructed to simply provide the best care possible, reported less concern for the pain of black patients and undertreated them in comparison to white patients (Drwecki et al. 2011). Imagine entering a hospital with a painful condition, desperate to feel better. But because someone can't imagine or feel your pain, you receive less relief and comprehensive treatment than you need, unnecessarily prolonging your physical suffering and emotional anguish. In short, empathy can matter as much as evidence when it comes to how we respond to problems. Sometimes it matters even more.

Empathy is not only important in these contexts but also debates over contested and racialized social issues. These debates rely heavily on identifying what and who matters and what we should do as a society and as individuals. Our relative capacity for empathy with others influences how we collectively define social problems and how we understand and respond to social problems as a society impacts our policies, institutions, and everyday lives. And in the heat of these debates, people construct and express racial identities that either fortify or weaken the racial empathy gap. Controlling images of black and Latinx people (Collins 2009), images that make their treatment in society seem a natural result of who they are, enabled many participants in the debate to ignore the full humanity of the people under consideration. These images are dangerous because they hold the power to allow us to forget that we are even talking about human beings. In many of the comments, because whites saw themselves as distinct, and in many cases superior in social or moral terms, from racial "others," they failed to recognize and value the social suffering caused to families and communities of color by unjust social arrangements and policies and practices like the War on Drugs.

Additionally, **moral categories**, often split into binaries like good and evil, innocent and guilty, or law abiders and criminals hold immense weight in our deliberations over social issues via our ability to empathize with others. We tend to empathize with those we see as innocent and not those that we see as criminal or guilty. In some ways, this distinction makes sense. But when our sense of guilt or innocence is distorted by associating these moral categories with racial categories rather than a clear understanding of power, inequality, and historical and social processes, we are plunged into a world where our capacity for empathy is limited by racial boundaries. Racial justice activist and public intellectual Malcolm X proclaimed the media as "the most powerful entity on earth" because they "have the power to make the innocent guilty and to make the guilty innocent" (as quoted in Griffith 2012:115). As we saw in newspaper contents on the War on Drugs, we are bombarded with images and stories in the media that confirm these correlations or ignore the issues of racial justice altogether.

So, alongside empathy, morality also matters. However, I don't mean morality in terms of some objective moral absolutes. As the 19th-century German philosopher Friedrich Wilhelm Nietzsche (1886[1907]:91) famously argued, "there is no such thing as moral phenomena, but only a moral interpretation of phenomena." From a sociological perspective, morality matters in how people use moral ideas, categories, and ideals to construct a sense of who they are in relation to the world around them and the other people that populate it. Ideas about morality play a crucial role in the construction of people's racial identities (Lamont 2000).

In Chapter 5, we saw the dominant role of morality in identity construction within the debate. And it mattered not just for racial identity construction but also for political identity construction. Consider how often claims of moral hypocrisy or inferiority were attributed to political categories such as conservatives or liberals throughout the debate. People's claims about who they were, in racial and political terms, were not just dependent on drawing symbolic boundaries but also often about placing themselves and their imagined group members in a sort of moral and social hierarchy with others.

Political and racial identities were often connected and reinforced through claims about the morality of the self and immorality of others. Yet political divides, while salient, were not total fault lines. We saw this in the shared racialized subject-positions that people often identified with in relation to "immoral" racialized others, regardless of their stance on the issue of drug policy itself. Divergent identities in terms of the position people took on the War on Drugs at times manifested as mere disagreements over how to best subdue or control those groups associated with racialized moral panics.

One of the major goals of this book has been to help clarify the relationship between debates in the media and **identities**. We have explored how people collectively define social problems and construct identities in debates over contested social issues. And we have uncovered connections between these processes. But perhaps just as important is, if not more so, understanding how collective and individual identities impact society. As Berger and Luckmann (1966:173) pointed out, "the identities produced by the interplay of organism [the physical person], individual consciousness, and social structure react upon the given social structure, maintaining it, modifying it, or even reshaping it." Our identities manifest in our thoughts, words, and deeds as we "react upon" the social world that we encounter daily. So, our identities, our sense of self, or who we are do not simply matter because they have social causes. They matter because they have vast social consequences.

Stuart Hall (1996:16) wrote that "the question, and the theorization, of identity is a matter of considerable political significance." Political conflicts often boil down to group conflicts over power. Power is a capacity for action determined by the social context, including the ability to influence social conditions (Rosino 2016). This gets at something more important and fundamental than the often-heard phrase "identity politics." Focusing on this political significance means seriously taking stock of the role of power in the relationship between identities and group relations. Examining debates over contested social issues helps us understand how people form their sense of who they are in ways that reinforce, redefine, and dispute not just social problems but also unjust social arrangements. Identities contribute to inequality, injustice, and suffering when they are produced through a sense that their social group's domination over others is natural, inevitable, and legitimate.

Prejudice and discrimination by whites toward black, Latinx, Asian, and Native people depend upon a sense of in-group superiority, a sense that others are fundamentally alien or different, a sense of entitlement to advantages and privileges based on group membership, and a sense that so-called "subordinate" racial groups threaten that superiority or dominant social position (Blumer 1958). Identities that reproduce social inequality involve dominant group members cultivating a sense of group position through a collective process in which they define their own racial group and other groups (Blumer 1958). Throughout this book, we see how participants in the debate argued against the War on Drugs, a harmful set of policies and practices, but left these elements of the sense of group position held by whites and the commonsense myths related to them unchallenged.

Our sense of who we are is reflected in what we do and say in our daily lives. And what we do and say matters when it comes to how we respond to the reality of racial oppression because we have agency. In Chapters 1 and 5, I mentioned that identities are shaped by the way society is arranged and the choices we make in interactions – that it is a matter of both structure and agency. Agency is not simply about what we intend to do, but rather about the specific actions we take given the range of possibilities (Giddens 1993). Mass media and other social institutions present us with racialized subject-positions, possible ways of seeing the world, and defining who we are in relation to others, but we have agency over whether to identify with them.

The social context influences our options. Some of us have more power and resources than others, but we all encounter a range of possibilities. While we have less control over the meanings and labels that we are ascribed by institutions and other people, we have more agency over our sense of who we are in relation to others. Patricia Hill Collins (2009) has called this use of agency the power of self-definition, and points out that it has long served as an important tool for oppressed groups to engage in struggles for social change and justice.

Acknowledging agency doesn't mean that people who are oppressed by unfair social arrangements are solely responsible for transforming them. Those of us with influence and access to resources have a unique responsibility to challenge dominant racial meanings and dismantle systems that maintain inequality. The choices we make in defining who we are depend on the **identity opportunities** we encounter or the range of possibilities we have available to identify and define ourselves. Collective and organized action can reshape and restructure these opportunities (see Omi and Winant 1994). We should all use our agency to advance narratives, categories, and ideas that help others build a sense of self that is rooted in sociological knowledge and opposition to injustice and inequality.

Looking at the debate on the War on Drugs, it's clear that **sociological knowledge**, an understanding of causes and consequences based on sociological thinking and empirical facts, needs to play a stronger role in the debates that impact policies and institutional arrangements. At the same time, this knowledge can also be a potent source of self-reflection that leads everyday people toward enacting social change. Identity is "strategic and positional" (Hall 1996:3) rather than fixed and essential. It is part of a constantly unfolding process of definition. Sociological knowledge about things like inequality, social processes, and shared meaning helps us align our intentions with our actions if we are willing to integrate it into how we make sense of ourselves, others, and the society that we share. Sociological knowledge must not just be *informative*; it must also be *transformative* as well.

It can be difficult to constantly act, think, and speak in ways that challenge forms of social inequality and the meanings that perpetuate them. Even if we cultivate the knowledge and empathy necessary, personal rewards and punishments are at times dependent on whether we conform to dominant expectations. The power of commonsense myths is drawn from the "sense that these beliefs are what 'most people' believe and thus are the public rules by which [we] will be judged and expected to act" (Ridgeway and Kricheli-Katz 2013:302). And once someone identifies with a set of racial meanings they can become dependent on them for their sense of self.

Identities can be laden with expectations about how we should think, talk, or act (Schwalbe 2008). This sense of social judgment and expectation reflects racial group differences in what sociologist Pierre Bourdieu (1991) calls **symbolic power**. Dominant groups impose categories and discourses that reflect their interests onto the social world as legitimate, natural, or commonsense. In predominantly white social settings, black people often struggle to disprove or define themselves against the stereotypes and myths held by whites as commonsense (Anderson 2016). And whites are often held accountable by other whites for preserving idealized interests whether through racial silence and inaction, employing coded language or implicit racial meanings, or accepting commonsense myths.

Yet, even if it holds some emotional and material rewards for the individual, when dominant group members define themselves and others with dominant forms of identity, the effects on agency and the capacity for empathy produces inequality and social suffering. All of us must be willing to reconsider who we are so that we can advance a society that displays and encourages empathy toward all people. Bringing about these social changes means changing our shared meanings, institutional policies, and everyday social interactions.

Implications for Media, Policy, and Structural Change

Through examining the debate over the War on Drugs, we also learned a lot about mass media. Mass media, including popular digital media, provide a space where various viewpoints over contested social issues can be found. Yet, oftentimes, mass media content amplifies certain messages that defend the status quo and delegitimizes those related to efforts to change it (Hall et al. 1971; Schwalbe 2008). In Chapter 3, we covered how the media has immense influence in setting the agenda and framing debates. These debates determine how we collectively define and approach our most pressing social problems.

We live in a media-saturated society (Couldry 2012). Dominant group influence over media content affects what we see and hear so we must critically examine the images, meanings, and messages that media content presents as natural, normal, or commonsense (Kellner and Share 2005; Luke 1994). As media consumers and audiences, we need better tools for contextualizing and interpreting the media images and stories. In other words, we need to develop what education and cultural studies scholars call **critical media literacy** (Luke 1994).

But as noted by Jessie Daniels (2008), a digital sociologist, online spaces present unique opportunities for racially segregated digital communities and racial discourse, and thus require a specific form of **digital racial literacy** for participants and researchers. Foreshadowing the recent explosion of Internet-based white supremacist movements, Daniels (2008:131) reminds us that "hate speech online (O.L.) can have very real consequences in real life (IRL)." Yet, it is not just overt hate speech online that matters. In more subtle ways, dominant racial meanings pervade mainstream and popular online spaces and shape the discussion of racialized social issues such as the War on Drugs. In participating in debates over racialized and contested social issues via mass media, especially digital media, we need to consider which perspectives represent reality and which are downplayed or silenced, especially in constructing contested social issues as having two sides.

Information circulated through the media plays an important role in many whites' perceptions of other racial groups and society. Matthew Robinson (2000), a criminologist, argues that the media reinforces myths about race and crime. In contrast with reality, media depictions suggest that most crime is violent; that most crime is committed by blacks; and that crime is pervasive; and therefore, the majority of Americans should be fearful (Robinson 2000). In network news reportage of crime, blacks are underrepresented as victims and police officers and overrepresented as perpetrators (Dixon, Azocar, and Casas 2003). Even the images utilized by national news during noncrime reporting disproportionately depict blacks as poor, aggressive, and criminal (Entman and Rojecki 2001). Looking at the War on Drugs debate in journalistic reporting and opinion writing in print media and digital media commenting among audiences demonstrates even further how images and discourses in mass media naturalize the criminalization and oppression of black and Latinx people. Moreover, it shows us that these images and discourses play a crucial part in the sense of innocence, victimization, and entitlement to a dominant status that many whites use to develop their identities.

The media amplifies and silences voices in the conversations that determine how we define social problems. When we see the media defining social issues as social problems, we must deeply consider who is most affected, who has power, and the full implications of proposed reactions or solutions. Educators and researchers have a responsibility to counterbalance the fact that the media:

> provide powerful public pedagogies which shape concepts of self, gender and race identity and relations; ideas about which social groups count as culturally relevant and politically powerful; and what counts as 'history,' 'progress,' 'science,' 'cultural difference,' 'family,' 'individuality,' or 'political processes.'
> (Luke 1994:31)

Yet, once again, our engagement with media and public debates needs to go beyond just myth debunking.

What often matters in our debates over contested social issues is the stories we hear and tell: characters and their relationships, big events, heroes, villains, hopes, fears, threats, failures, tragedies, and victories. It matters whose perspective the story is being told from, who is the "we" or "us," and who is the "they" or "them," and what the boundaries between these categories mean. If we pay close attention to these stories and their elements, we can better participate in debates about contested social issues and help steer them toward solutions that combat racial oppression, recognize and reduce social suffering, and bring about a more just and empathetic society.

The debate about the War on Drugs also tells us about how people relate to policies and how policies affect people. People are increasingly dissatisfied with US drug policies and how they are enforced. Support for the legalization of cannabis has been steadily rising since the early 1990s and a majority of Americans now agree that the substance should be legal (Pew Research 2016b). Contemporary public opinion polling also suggests higher support for treatment than punishment as a response to drug abuse (Pew Research 2014).

Yet, as long as the narratives and categories that fundamentally equate nonwhites with threat or dysfunction remain dominant within both sides of the debate, policy reform efforts will fail to fully address the tragic and oppressive racial legacy of the War on Drugs. For instance, while the legalization of cannabis in Colorado has reduced possession arrests overall, vast racial disparities in arrests for public consumption or underage possession remain (Drug Policy Alliance 2015; NPR 2016). Similarly, in 2019, as Illinois joined other states in legalizing recreational cannabis use for adults, the disproportionately poor and nonwhite tenants of public housing were informed that they could still face eviction for its possession and use, even for those who have a relevant medical condition (Romo 2019). And the legal cannabis industry has remained overwhelmingly white due to legal and social barriers to entry, and so the profits and legitimacy of this emerging market have passed over those who faced the most costs from prohibition (Lewis 2016). These remaining inequities are, in a large part, the products of racial silence in the debate. The process through which we come to define social problems, especially the question of why something is problematic, matters in crafting public policies and other solutions that can address social problems and their consequences.

Perhaps most importantly of all, a major goal of this book is to provide fodder for considering how things like media, policies, identities, language, institutions, and debates influence structural change. **Structural change** means transforming how rewards, punishments, and opportunities are distributed among groups and individuals. It also means altering the informal and formal rules that govern how we interact with each other. Structural change is not an anomaly. It is a constant. When we look at the history of our society, we have plenty of examples of how powerful groups have encouraged structural changes to maintain social inequality and block social progress (Anderson 2016).

The growth of racialized mass incarceration in the shadow of chattel slavery and Jim Crow is an example of how the structure of society can change and yet still function to reproduce racial inequality. However, structural change is the product of the collective choices people make. It is not something that naturally takes place. So, the question at hand is not one of change versus stagnation, but rather how we can direct structural change toward the alleviation of social problems and injustice rather than the maintenance of systems and practices of oppression.

Throughout this book, I have introduced and detailed an abundance of problems relating to racial oppression and the War on Drugs such as warrior-style policing, the militarization of the police, racial discrimination in the legal system, racial inequality in wealth, and controlling images and representations that dehumanize black and Latinx people. In one sense, all this is troubling. Yet in another sense, these varied problems suggest a multitude of opportunities to get involved in efforts to bring about meaningful change.

Take racialized mass incarceration for example. Drug prohibition laws in the United States have played a huge role in the production and protection of this system (Cooper 2015; Rosino and Hughey 2018). Alongside legal reforms that legalize drug use, we also need to repair the social damage that our drug laws have wrought on families and communities. The beneficiaries of the new economic and social opportunities that come from drug reforms must be those who suffered most as the drug war's victims. For instance, San Francisco's Cannabis Equity Program waives permit fees and facilitates licensing for cannabis sellers "who meet equity eligibility criteria based on residency, income, criminal justice involvement, and housing insecurity" (San Francisco Office of Cannabis 2019).

The larger goal beyond legalization and reparations is decarceration. **Decarceration** means reducing the jail and prison population, reducing the use of incarceration to address social problems, and reforming the legal system overall toward these ends (Drucker 2017). Decarceration also requires institutional changes that improve people's ability to gain education, healthcare, and economic stability and empower communities to use restorative justice and rehabilitation rather than punishment. It is not simply enough to remove mass incarceration, but rather to replace it with a new set of practices and systems that uphold human dignity while dealing with the very real problems of violence, victimization, and injustice by identifying and addressing their underlying causes (Drucker 2017).

Affecting structural change is a big task but its immensity is not a cause for inaction. You can work towards structural change wherever you are. Take some time to think deeply about who you are. Try to place your own sense of self in a historical and social context. As you navigate everyday life, consider all the ways that social arrangements affect people and don't settle for explanations that blame the victim. When you engage in conversations, especially those about racialized and contested social issues, listen to the stories and claims of others with an open mind but think critically about the dynamics of power and social control.

Public debates over social issues are not just intellectual exercises or competitions. They translate directly to the real world. They impact the collective ways that we define and solve problems that impact real people. However, as we can see in the War on Drugs debate, a false sense of neutrality is routinely weaponized by dominant groups to avoid acknowledging injustice. Moreover, our engagement in debates can often serve as a way to express our sense of social and political identity rather than a means of productive engagement and mutual understanding. Rather than avoiding the reality of social inequalities and group interests or merely implying them, the health of our society depends on having public conversations that critically examine them.

Just rethinking how you engage with others in everyday life can make a positive impact on our world. However, to affect structural change, people need to be intentional, active, and organized. Form coalitions and alliances with those who have shared goals in addressing social problems. For instance, the history of social movements demonstrates that multiracial organizations, especially those in which black, Latinx, Asian, and Native people have a strong voice, if not leadership, can be effective in combatting racial oppression. Remember that social problems have social solutions, not individual ones. In other words, if you want to work for structural change, don't go it alone. If you care about social problems and inequality, know that others care as well, so you don't need to start from scratch. Seek out organizations and groups that already exist and could use your help.

For your convenience, I've listed some here:

Drug Policy and Prison Reform

- Drug Policy Alliance (drugpolicy.org)
- Students for Sensible Drug Policy (ssdp.org)
- The Sentencing Project (sentencingproject.org)
- Prison Policy Initiative (prisonpolicy.org)
- Equal Justice Initiative (eji.org)

Legal, Civil, and Human Rights

- American Civil Liberties Union (aclu.org)
- National Lawyers Guild (nlg.org)

- Human Rights Watch (hrw.org)
- U.S. Human Rights Network (ushrnetwork.org)
- Southern Poverty Law Center (splc.org)
- NAACP Legal Defense and Educational Fund (naacpldf.org)

Media Literacy, Freedom, and Equity

- Free Press (freepress.net)
- Center for Media Justice (centerformediajustice.org)
- Center for Media Literacy (medialit.org)
- Center for Digital Democracy (democraticmedia.org)

Racial Justice and Liberation from Oppression

- Center for Community Change (communitychange.org)
- Black Youth Project 100 (byp100.org)
- Race Forward (raceforward.org)
- Color of Change (colorofchange.org)
- The Movement for Black Lives (policy.m4bl.org)

Discussion Questions

1 What were some things that you know now that you did not know about society, yourself, the media, politics, racial inequality, or other topics that you now know? What are some remaining questions you have? How might you go about finding answers to those questions?

2 Can you think of a time that you encountered a commonsense myth concerning the War on Drugs debate? How did you respond? How did it make you feel? What are some ways to challenge these myths or reduce their influence?

3 Think of some people that you feel a sense of empathy toward. How would you describe that feeling and the ways that it makes you think and act? What can we do as a society to ensure that we collectively feel that sense toward people facing unfair social arrangements? What are some steps that people could take to feel empathy across symbolic boundaries?

4 Think about some media content that you consume on a daily or regular basis. What kinds of people are represented? What are the types of storylines depicted? Can you think of any ways

that these representations and storylines might affect your sense of who you are or how you think, feel, and act in everyday life?
5 Pick one issue or problem discussed in this book that most stood out to you. What are the causes and consequences of it? What are some concrete steps you will take to put your knowledge of these causes and consequences into action?

Notes

1 See Haney López (2014) on "commonsense racism."
2 See Haney López (2014) on "strategic racism."
3 See Silverstein (2013) who notes "the racial empathy gap helps explain disparities in everything from pain management to the criminal justice system. But the problem isn't just that people disregard the pain of black people. It's somehow even worse. The problem is that the pain isn't even felt." And Bonilla-Silva (2016:245) writes, "for whites to join people of color all the way in the struggle, they must regard them in a profound way as their equals; they must be able to empathize with them."

References

Ahmad, Diana L. 2000. "Opium Smoking, Anti-Chinese Attitudes, and the American Medical Community, 1850–1890". *American Nineteenth Century History* 1(2):53–68.
Alexander, Michelle. 2012. *The New Jim Crow: Mass Incarceration in the Age of Colorblindness.* New York, NY: The New Press.
Allen, Theodore W. 1997. *The Invention of the White Race: Volume Two, The Origins of Racial Oppression in Anglo-America.* New York, NY: Verso Books.
Altheide, David L. and Christopher J. Shneider. 2013. *Qualitative Media Analysis.* 2nd Edition. New York, NY: Sage.
Althusser, Louis. [1971] 2001. *Lenin and Philosophy and Other Essays.* New York, NY: Monthly Review Press.
American Journal of Bioethics. Forthcoming. "Racial Justice Requires Drug Policy Overhaul: Bioethicists and Others against the War on Drugs". *American Journal of Bioethics.*
Anderson, Carol. 2016. *White Rage: The Unspoken Truth of Our Racial Divide.* New York, NY: Bloomsbury.
Angwin, Julia, Surya Mattu, and Lauren Krichner. 2016. "Machine Bias". ProPublica May 23. Located online: https://www.propublica.org/article/machine-bias-risk-assessments-in-criminal-sentencing
Armaline, William T. 2011. "Caging Kids of Color: Juvenile Justice and Human Rights in the United States". Pp. 189–198 in *Human Rights in Our Own Backyard.* Edited by William Armaline, Davita Glasberg, and Bandana Purkayastha. Philadelphia, PA: University of Pennsylvania Press.
Bagdikian, Ben H. 1983. *The Media Monopoly.* Boston, MA: Beacon Press.
Balko, Radley. 2013. *Rise of the Warrior Cop: The Militarization of America's Police Forces.* New York, NY: PublicAffairs.
Barbaro, Michael. 2016. "Donald Trump Clung to 'Birther' Lie for Years, and Still Isn't Apologetic". *The New York Times* Sep 16. Accessed: https://www.nytimes.com/2016/09/17/us/politics/donald-trump-obama-birther.html.

Barthes, Roland. 1957. *Mythologies*. Translated by Annette Lavers. New York, NY: The Noonday Press.

Baum, Dan. 1996. *Smoke and Mirrors: The War on Drugs and the Politics of Failure*. Boston, MA: Little, Brown, and Company.

Becker, Howard S. 1963. *Outsiders: Studies in the Sociology of Deviance*. New York, NY: Free Press.

Beckett, Katherine. 1994. "Setting the Public Agenda: "Street Crime" and Drug Use in American Politics". *Social Problems* 41(3):425–447.

Beckett, Katherine, Kris Nyrop, and Lori Pfingst. 2006. "Race, Drugs, and Policing: Understanding Disparities in Drug Delivery Arrests". *Criminology* 44(1):105–137.

Beckett, Katherine, Kris Nyrop, Lori Pfingst, and Melissa Bowen. 2005. "Drug Use, Drug Possession Arrests, and the Question of Race: Lessons from Seattle". *Social Problems* 52(3):419–441.

Berensen, Alex. 2019. *Tell Your Children: The Truth about Marijuana, Mental Illness, and Violence*. New York, NY: Free Press.

Berger, Peter L. and Thomas Luckmann. 1966. *The Social Construction of Reality: A Treatise in the Sociology of Knowledge*. New York, NY: Penguin Books.

Blauner, Robert. 1972. *Racial Oppression in America*. New York, NY: HarperCollins.

Blumer, Herbert. 1958. "Race Prejudice as a Sense of Group Position". *The Pacific Sociological Review* 1(1):3–7.

Blumer, Herbert. 1971. "Social Problems as Collective Behavior". *Social Problems* 18(3):298–306.

Bonilla-Silva, Eduardo. 1997. "Rethinking Racism: Toward a Structural Interpretation." *American Sociological Review* 62(3):465–480.

Bonilla-Silva, Eduardo. 1999. "The Essential Social Fact of Race". *American Sociological Review* 64(6):899–906.

Bonilla-Silva, Eduardo. 2002. "The Linguistics of Color Blind Racism: How to Talk Nasty about Blacks without Sounding "Racist"". *Critical Sociology* 28(1–2):41–64.

Bonilla-Silva, Eduardo. 2006. "From Bi-racial to Tri-racial: Towards a New System of Racial Stratification in the USA". *Ethnic and Racial Studies* 27(6):931–950.

Bonilla-Silva, Eduardo, and David G. Embrick. 2007. "'Every Place Has a Ghetto...': The Significance of Whites' Social and Residential Segregation". *Symbolic Interaction* 30(3):323–345.

Bonilla-Silva, Eduardo. 2014. *Racism without Racists: Color-Blind Racism and the Persistence of Racial Inequality in Contemporary America*. 4th Edition. Lanham, MD: Rowman and Littlefield Publishers, Inc.

Bonilla-Silva, Eduardo. 2016. "Reply to Professor Fenelon and Adding Emotion to My Materialist RSS Theory". *Sociology of Race and Ethnicity* 2(2):243–247.

Bourdieu, Pierre. 1991. *Language and Symbolic Power*. Cambridge, UK: Polity Press.
Bourdieu, Pierre. 1998. *Practical Reason: On the Theory of Action*. Stanford, CA: Stanford University Press.
Brantley, Shartia. 2011. "Who Are the Black '1 percent'?". *The Grio* Nov 21. Accessed: http://thegrio.com/2011/11/21/who-are-the-black-1-percent/.
Bredderman, Will. 2016. "NYPD Reports 'Huge Spike' in Hate Crimes Since Donald Trump's Election". Observer, Dec 5. Located online: https://observer.com/2016/12/nypd-reports-huge-spike-in-hate-crimes-since-donald-trumps-election/
Britton, Tolani. 2019. "Does Locked Up Mean Locked Out? The Effects of the Anti-Drug Act of 1986 on Black Male Students' College Enrollment". IRLE Working Paper #101–19.
Brown, Hanna E. 2013. "Racialized Conflict and Policy Spillover Effects: The Role of Race in the Contemporary U.S. Welfare State". *American Journal of Sociology* 119(2):349–443.
Bruenig, Matt. 2014. "The Top 10% of White Families Own Almost Everything". *Demos* Sep 5. Accessed: http://www.demos.org/blog/9/5/14/top-10-white-families-own-almost- everything.
Butcher, Kristen F. and Anne Morrison Piehl. 2007. "Why are Immigrants' Incarceration Rates so Low? Evidence on Selective Immigration, Deterrence, and Deportation". NBER Working Paper 13229.
Byrd, W. Carson and Matthew W. Hughey. 2015. "Biological Determinism and Racial Essentialism: The Ideological Double Helix of Racial Inequality". *The ANNALS of the American Academy of Political and Social Science* 661(1):8–22.
Byrd, W. Carson and Victor E. Ray. 2015. "Ultimate Attribution in the Genetic Era: White Support for Genetic Explanations of Racial Difference and Policies". *The ANNALS of the American Academy of Political and Social Science* 661(1):212–235.
Case, Anne C. and Lawrence F. Katz. 1991. "The Company You Keep: The Effects of Family and Neighborhood on Disadvantaged Youths". *NBER Working Papers*. Accessed: http://www.nber.org/papers/w3705.pdf.
Cheatwood, Derral. 2010. "Images of Crime and Justice in Early Commercial Radio – 1932 to 1985". *Criminal Justice Review* 35(1):32–51.
Clear, Todd R. 2007. *Imprisoning Communities: How Mass Incarceration Makes Disadvantaged Neighborhoods Worse*. Oxford, UK: Oxford University Press.
Clegg, John and Adaner Usmani. 2019. "The Economic Origins of Mass Incarceration." *Catalyst* 3(3). Located online at: https://catalyst-journal.com/vol3/no3/the-economic-origins-of-mass-incarceration
Clinard, Marshall B. and Robert F. Meier. 2001. *Sociology of Deviant Behavior*. San Diego, CA: Harcourt College Publishers.

Coates, Rodney D. 2003a. "Introduction: Reproducing Racialized Systems of Social Control". *American Behavioral Scientist* 47(3):235–239.

Coates, Rodney D. 2003b. "Law and the Cultural Production of Race and Racialized Systems of Oppression: Early American Court Cases". *American Behavioral Scientist* 47(3):329–351.

Cohen, Michael M. 2006. "Jim Crow's Drug War: Race, Coca Cola, and the Southern Origins of Drug Prohibition". *Southern Cultures* 12(3):55–79.

Cohen, Stanley. 1972. *Folk Devils and Moral Panics: The Creation of the Mods and Rockers*. London, UK: MacGibbon and Kee Ltd.

Collins, Allyson. 1998. *Shielded from Justice: Police Brutality and Accountability in the United States*. New York, NY: Human Rights Watch.

Collins, Patricia Hill. 1993. "Toward a New Vision: Race, Class, and Gender as Categories of Analysis and Connection". *Race, Sex & Class* 1(1):25–45.

Collins, Patricia Hill. 2009. *Black Feminist Thought*. New York, NY: Routledge.

Connell, R. W. 1995. *Masculinities*. Berkeley: University of California Press.

Cooley, Charles Horton. 1906. *Human Nature and the Social Order*. New York, NY: Shocken.

Coontz, Stephanie. 1992. *The Way We Never Were: American Families and the Nostalgia Trap*. New York, NY: Basic Books.

Cooper, Hannah L. F. 2015. "War on Drugs Policing and Police Brutality". *Substance Use & Misuse* 50(8–9):1188–1194.

Cortright, David. 2005. *Soldiers in Revolt: GI Resistance during the Vietnam War*. New York, NY: Haymarket Books.

Couldry, Nick. 2012. *Media, Society, World: Social Theory and Digital Media Practice*. New York, NY: Polity Press.

Dalton, Amy N. and Li Huang. 2014. "Motivated Forgetting in Response to Social Identity Threat". *Journal of Consumer Research* 40(6):1017–1038.

Daniels, Jessie. 2008. "Race, Civil Rights, and Hate Speech in the Digital Era". Pp. 129–154 in *Learning Race and Ethnicity: Youth and Digital Media*. Edited by Anna Everett. Cambridge, MA: The MIT Press.

Daniels, Jessie. 2013. "Race and Racism in Internet Studies: A Review and Critique". *New Media & Society* 15(5):695–719.

Davenport, Sarah. 2011. "1971: Forty Years since Nixon's 'War on Drugs'". *The Guardian* Jul 22. Accessed: https://www.theguardian.com/theguardian/from-the-archive-blog/2011/jul/22/drugs-trade-richard-nixon.

Davies, Bronwyn and Rom Harré. 1990. "Positioning: The Discursive Construction of Selves". *Journal for the Theory of Social Behavior* 20(1):43–63.

Deetz, Stanley A., Sarah J. Tracy and Jennifer Lyn Simpson. 2000. *Leading Organizations through Transition: Communication and Cultural Change*. New York, NY: Sage.

Delgado, Richard. 2000. "Storytelling for Oppositionists and Others: A Plea for Narrative". Pp. 60–70 in *Critical Race Theory: The Cutting Edge*. 2nd Edition. Edited by Richard Delgado and Jean Stefancic. Philadelphia, PA: Temple University Press.

Democracy Now!. 2012. "Ramarley Graham, Unarmed Black Teen Slain by NYPD, Remembered at Weekly Vigils Outside Bronx Home". Jun 19. Accessed: https://www.democracynow.org/2012/6/19/ramarley_graham_unarmed_black_teen_slain.

Denvir, Daniel. 2015. "Violent Crime Rates: Still Declining". *CityLab* Sep 29. Accessed: http://www.citylab.com/crime/2015/09/violent-crime-rates-still- declining/408103/.

Desrochers, Jr., Robert E. 2002. "Slave-for-Sale Advertisements and Slavery in Massachusetts, 1704–1781". *The William and Mary Quarterly* 59(3):623–664.

Dixon, Travis L. 2007. "Black Criminals and White Officers: The Effects of Racially Misrepresenting Law Breakers and Law Defenders on Television News." *Media Psychology* 10:270–291.

Dixon, Travis L., Cristina L. Azocar, and Michael Casas. 2003. "The Portrayal of Race and Crime on Television Networks". *Journal of Broadcasting & Electronic Media* 47(4):498–523.

Doane, Ashley W., Jr. 2003. "Contested Terrain: Negotiating Racial Understandings in Public Discourse". *Humanity & Society* 27(4):554–575.

Doane, Ashley W. 2006. "What Is Racism? Racial Discourse and Racial Politics". *Critical Sociology* 32(2–3):255–274.

Doering, Jan. 2014. "A Battleground of Identity: Racial Formation and the African American Discourse on Interracial Marriage". *Social Problems* 61(4):559–575.

Downing, John D. H. and Charles Husband. 2005. *Representing Race: Racisms, Ethnicities, and Media*. CA: Sage Publications.

Dropp, Kyle and Brendan Nyhan. 2016. "It Lives. Birtherism Is Diminished but Far from Dead." *The New York Times*. Accessed: https://www.nytimes.com/2016/09/24/upshot/it-lives-birtherism-is-diminished-but-far- from-dead.html.

Drucker, Ernest. 2017. "A Public Health Approach to Decarceration: Strategies to Reduce the Prison and Jail Population and Support Reentry". Pp. 179–192 in *Smart Decarceration: Achieving Criminal Justice Transformation in the 21st Century*. Edited by M. Epperson and C. Pettus Davis. New York: Oxford University Press.

Drug Policy Alliance. 2015. "Marijuana Arrests in Colorado after the Passage of Amendment 64". Report Prepared by Jon Gettman, Mar

25. Accessed: http://www.drugpolicy.org/sites/default/files/Colorado_Marijuana_Arrests_After_Amendment_64.pdf.
Drug Policy Alliance. 2017. "Wasted Tax Dollars". Accessed: http://www.drugpolicy.org/wasted-tax-dollars.
Drug Policy Alliance. 2019. "Letter from Scholars and Clinicians Who Oppose Junk Science about Marijuana." Feb 14. Accessed: https://www.drugpolicy.org/resource/letter-scholars-and-clinicians-who-oppose-junk-science-about-marijuana.
Drwecki, Brian B., Colleen F. Moore, Sandra E. Ward, Kenneth M. Prkachin. 2011. "Reducing Racial Disparities in Pain Treatment: The Role of Empathy and Perspective-Taking". *Pain* 152(2):1001–1006.
Du Bois, William Edward Burghardt. 1888. *The Philadelphia Negro: A Social Study*. Philadelphia, PA: University of Pennsylvania Press.
Du Bois, William Edward Burghardt. 1889. "The Study of the Negro Problem". *Annals of the American Academy of Political and Social Sciences* 1:1–23.
Du Bois, William Edward Burghardt. 1935. *Black Reconstruction in America*. New York, NY: Simon & Schuster.
Durkheim, Emile. [1893] 1984. *The Division of Labor in Society*. Translated by W. D. Halls. New York, NY: The Free Press.
Durkheim, Emile. [1912] 1995. *The Elementary Forms of Religious Life*. Translated by Karen E. Fields. New York, NY: The Free Press.
Ellwood, C. A. 1912. "Immigration and Crime". *Journal of the American Institute of Criminal Law and Criminology* 3(1):8–10.
Elwood, William N. 1994. *Rhetoric on the War on Drugs: The Triumphs and Tragedies of Public Relations*. Westport, CT: Praeger.
Embrick, David G. and Kasey Henricks. 2013. "Discursive Colorlines at Work: How Epithets and Stereotypes are Racially Unequal". *Symbolic Interaction* 36(2):197–215.
Emerson, Blake. 2011. "Criminal Justice and the Ideology of Individual Responsibility". Pp. 65–78 in *Race, Crime, and Punishment: Breaking the Connection in America*. Edited by Keith O. Lawrence. Washington, DC: The Aspen Institute.
Entman, Robert and Andrew Rojecki. 2001. *The Black Image in the White Mind: Media and Race in America*. Chicago, IL: University of Chicago Press.
Epp, Charles R., Steven Maynard-Moody, and Donald P. Haider-Merkel. 2013. *Pulled over: How Police Stops Define Race and Citizenship*. Chicago, IL: University of Chicago Press.
Essed, Philomena. 1991. *Understanding Everyday Racism: An Interdisciplinary Theory*. Newbury Park, CA: Sage.
Fairhurst, Gail T. and Robert A. Star. 1996. *The Art of Framing: Managing the Language of Leadership*. New York, NY: Jossey-Bass.

Fairlie, Robert W. 2002. "Drug Dealing and Legitimate Self-Employment". *Journal of Labor Economics* 20(3):538–567.
Feagin, Joe R. 2001. *Racist America: Roots, Current Realities, and Future Reparations.* New York, NY: Routledge.
Feagin, Joe R. 2006. *Systemic Racism: A Theory of Oppression.* New York, NY: Routledge.
Feagin, Joe R. 2009. *The White Racial Frame: Centuries of Racial Framing and Counter-Framing.* New York, NY: Routledge.
Fellner, Jamie. 2009. "Race, Drugs, and Law Enforcement in the United States". *Stanford Law and Policy Review* 20(2):257–292.
FindLaw. 2013. "'Three Strikes' Sentencing Laws". *FindLaw.* Accessed: http://files.findlaw.com/pdf/criminal/criminal.findlaw.com_criminal-procedure_three-strikes-sentencing-laws.pdf.
Finkelman, Paul. 2012. "The Monster of Monticello." *New York Times* Nov 30. Accessed: http://www.nytimes.com/2012/12/01/opinion/the-real-thomas-jefferson.html.
Fisher, George. 2014. "The Drug War at 100". *Stanford Lawyer.* Accessed: https://law.stanford.edu/2014/12/19/the-drug-war-at-100/.
Flegenheimer, Matt and Al Baker. 2012. "Officer Fatally Shoots Teenager in Bronx". *The New York Times.* Feb 2. Accessed: http://www.nytimes.com/2012/02/03/nyregion/unarmed-teenager-fatally-shot-by-officer-chasing-him.html?_r=0.
Forman, Tyrone A. and Amanda E. Lewis. 2006. "Racial Apathy and Hurricane Katrina: The Social Anatomy of Prejudice in the Post-Civil Rights Era". *Du Bois Review* (3)1:175–202.
Foucault, Michel. 1971. *The Archeology of Knowledge and the Discourse on Language.* Translated by A. M. Sheridan Smith. New York, NY: Pantheon Books.
Fox, Cybelle. 2012. *Three Worlds of Relief: Race, Immigration, and the American Welfare State from the Progressive Era to the New Deal.* Princeton, NJ: Princeton University Press.
Frankenberg, Ruth. 1993. *White Women, Race Matters: The Social Construction of Whiteness.* Minneapolis: University of Minnesota Press.
Fraser, Nancy. 1992. "Rethinking the Public Sphere: A Contribution to the Critique of Actually Existing Democracy". Pp. 109–142 in *Habermas and the Public Sphere.* Edited by Craig Calhoun. Boston, MA: MIT Press.
Frick, Karin, Peter Gloor, and Detlef Gürtler. 2013. "Global-Thought-Leader 2013". *GDI Impuls* 4. Accessed: https://www.gdi.ch/media/News/Global_Thought_Leader_1_EN.pdf
Gabbidon, Shaun L. 2007. *W.E.B. Du Bois on Crime and Justice: Laying the Foundations of Sociological Criminology.* Farnham, UK: Ashgate Publishing.

Galliher, John F., David P. Keys, and Michael Elsner. 1998. "Lindesmith v. Anslinger: An Early Government Victory in the Failed War on Drugs". *The Journal of Criminal Law and Criminology* 88(2):661–682.

Garland, Tammy S. and Victor W. Bumphus. 2012. "Race, Bias, and Attitudes toward Drug Policy." *Journal of Ethnicity and Criminal Justice* 10:148–161.

Garner, Anne. 2014. "Chinese Opium Dens and the "Satellite Fiends of the Joints"". *New York Academy of Medicine Center for History* Oct 13. Accessed: https://nyamcenterforhistory.org/2014/10/13/chinatowns-opium-dens-and-the-satellite-fiends-of-the-joints/.

Gaytán, Marie Sarita. 2013. "Drinking Difference: Race, Consumption, and Alcohol Prohibition in Mexico and the United States". *Ethnicities* 14(3):436–457.

Giddens, Anthony. 1991. *Modernity and Self-Identity: Self and Society in the Late Modern Age*. Stanford, CA: Stanford University Press.

Giddens, Anthony. 1993. "Problems of Action and Structure". Pp. 88–175 in *The Giddens Reader*. Edited by Philip Cassell. Stanford, NJ: Stanford University Press.

Gilens, Martin. 1999. *Why Americans Hate Welfare: Race, Media, and the Politics of Antipoverty Policy*. Chicago, IL: University of Chicago Press.

Glenn, Evelyn Nakano. 2002. *Unequal Freedom: How Race and Gender Shaped American Citizenship and Labor*. Cambridge, MA: Harvard University Press.

Glickman, Lawrence B. 2018. "The Racist Politics of the English Language." *Boston Review of Books*, Nov 26. Accessed: http://bostonreview.net/race/lawrence-glickman-racially-tinged.

Goffman, Erving. 1959. *The Presentation of Self in Everyday Life*. New York, NY: Anchor Books.

Goffman, Erving. 1967. *Interaction Ritual: Essays on Face-to-Face Behavior*. New York, NY: Anchor Books.

Goffman, Erving. 1974. *Frame Analysis: An Essay on the Organization of Experience*. Boston, MA: Northeastern University Press.

Goldstein, Joseph. 2012. "Police Unit Faces Scrutiny after Fatal Shooting in the Bronx". *The New York Times* Feb 22. Accessed: http://www.nytimes.com/2012/02/23/nyregion/police-unit-faces-scrutiny-after-ramarley-grahams-death-in-the-bronx.html.

Goode, Erich. 2008. *Drugs in American Society*. New York, NY: McGraw-Hill Publishing.

Gossett, Thomas F. [1965] 1997. *Race: The History of an Idea in America*. New York, NY: Oxford University Press.

Gramsci, Antonio. [1929–1935] 1971. *Selections from the Prison Notebooks*. New York, NY: International Publishers.

Graves, Joseph L. 2001. *The Emperor's New Clothes: Biological Theories of Race at the Millennium*. New Brunswick, NJ: Rutgers University Press.

Greenbaum, Susan D. 2015. *Blaming the Poor: The Long Shadow of the Moynihan Report on Cruel Images about Poverty*. New Brunswick, NJ: Rutgers University Press.

Griffith, Joanne (Ed.). 2012. *Redefining Black Power: Reflections on the State of Black America*. San Francisco, CA: City Lights Books.

Gropper, Sareen S. and Jack L. Smith. 2013. *Advanced Nutrition and Human Metabolism*. 6th Edition. Belmont, CA: Wadsworth Publishing.

Habermas, Jurgen. [1962] 1989. *The Structural Transformation of the Public Sphere*. Boston, MA: MIT Press.

Habermas, Jurgen. [1964] 1974. "The Public Sphere: An Encyclopedia Article". Translated by Sara Lennox and Frank Lennox. *New German Critique* 3:49–55.

Hacker, Andrew. 1983. "Our Ministry of Information". *New York Times* Jun 26. Accessed: https://www.nytimes.com/1983/06/26/books/our-ministry-of-information.html

Hall, Stuart. 1980. "Encoding/Decoding". Pp. 117–127 in *Culture, Media, Language: Working Papers in Cultural Studies 1972–1979*. Edited by Stuart Hall, Dorothy Hobson, Andrew Lowe, and Paul Willis. New York, NY: Routledge.

Hall, Stuart. 1995. "The Whites of Their Eyes: Racist Ideologies and the Media". Pp. 18–22 in *Gender, Race, and Class in Media: A Text-Reader*. Edited by Gail Dines and Jean M. Humez. Thousand Oaks, CA: Sage.

Hall, Stuart. 1996. "Introduction: Who Needs Identity?". Pp. 1–17 in *Questions of Cultural Identity*. Edited by Stuart Hall and Paul Du Gay. London: Sage.

Hall, Stuart. 1997. "The Work of Representation". Pp. 13–74 in *Representation: Cultural Representations and Signifying Practices*. Edited by Stuart Hall. London: Sage.

Hall, Stuart, Chas Critcher, Tony Jefferson, John Clarke, and Brian Roberts. 1978. *Policing the Crisis: Mugging, The State, and Law and Order*. London: Macmillan Press.

Haney López, Ian F. 2006. *White by Law: The Legal Construction of Race*. New York, NY: New York University Press.

Haney López, Ian F. 2007. "Post-Racial Racism: Racial Stratification and Mass Incarceration in the Age of Obama". *California Law Review* 98:1023–1073.

Haney López, Ian F. 2014. *Dog Whistle Politics: How Coded Racial Appeals Have Reinvented Racism and Wrecked the Middle Class*. Oxford, UK: Oxford University Press.

Hari, Johann. 2015. "Philip Seymour Hoffman Could Be Alive Today If The Drug War Were Over". *The Huffington Post*. Accessed: http://www.huffingtonpost.com/johann-hari/philip-seymour-hoffman-wo_b_6576448.html

Hart, Carl L. 2013a. *High Price: A Neuroscientist's Journey of Self-Discovery That Challenges Everything You Know about Drugs and Society*. New York, NY: HarperCollins Publishers.

Hart, Carl L. 2013b. "Pot Reform's Race Problem". *The Nation* Nov 18: 17–18.

Hart, Carl L. 2017. "The Real Opioid Emergency". *The New York Times* Aug 18. Accessed: https://www.nytimes.com/2017/08/18/opinion/sunday/opioids-drugs-race-treatment.html.

Hauser, Christine. 2020. "Merriam-Webster Revises 'Racism' Entry after Missouri Woman Asks for Changes". Jun 10, *New York Times*. Accessed: https://www.nytimes.com/2020/06/10/us/merriam-webster-racism-definition.html.

Helmer, John and Thomas Vietorisz. 1974. *Drug Use, the Labor Market and Class Conflict*. Washington, DC: Drug Abuse Council.

Henricks, Kasey. 2017. ""I'm Principled against Slavery, but…": Colorblindness and the Three-Fifths Debate". *Social Problems* 65(3): 285–304. DOI: 10.1093/socpro/spx018.

Herer, Jack. 1998. *The Emperor Wears No Clothes: The Authoritative Historical Record of Cannabis and the Conspiracy against Marijuana*. Austin, TX: Ah Ha Publishing Company.

Hochschild, Jennifer. 1999. "Affirmative Action as Culture War". Pp. 343–368 in *The Cultural Territories of Race: Black and White Boundaries*. Chicago, IL: University of Chicago Press.

Hodges, Graham Russell and Alan Edward Brown. 1994. *"Pretends to be Free": Runaway Slave Advertisements from Colonial and Revolutionary New York and New Jersey*. New York, NY: Taylor & Francis.

Hourwich, Isaac A. 1912. "Immigration and Crime". *American Journal of Sociology* 17(4):478–490.

Howard, Judith A. 2000. "Social Psychology of Identities". *Annual Review of Sociology* 26:367–393.

Hughey, Matthew W. 2009. "Cinethetic Racism: White Redemption and Black Stereotypes in "Magical Negro" Films." *Social Problems* 56(3):543–577.

Hughey, Matthew W. 2010. "The White Savior Film and Reviewers' Reception". *Symbolic Interaction* 33(3):475–496.

Hughey, Matthew W. 2011. "Backstage Discourse and the Reproduction of White Masculinities". *Sociological Quarterly* 52(1):132–153.

Hughey, Matthew W. 2012a. *White Bound: Nationalists, Antiracists, and the Shared Meanings of Race*. Stanford, CA: Stanford University Press.

Hughey, Matthew W. 2012b. "'Show Me Your Papers!' Obama's Birth and the Whiteness of Belonging". *Qualitative Sociology* 35(2):163–181.

Hughey, Matthew W. 2014. "White Backlash in the 'Post-Racial' United States". *Ethnic and Racial Studies* 37(5):721–730.

Hughey, Matthew W. and Jessie Daniels. 2013. "Racist Comments at Online News Sites: A Methodological Dilemma for Discourse Analysis". *Media, Culture & Society* 35(3):332–347.

Hughey, Matthew W. and Gregory S. Parks. 2014. *Wrongs of the Right: Language, Race, and the Republican Party in the Age of Obama*. New York: NYU Press.

Hughey, Matthew W. and Michael L. Rosino. Forthcoming. "Make America White Again: The Racial Reasoning of American Nationalism." *Structural Racism and the Root Causes of Prejudice*. Edited by H. Mahmoudi and R. Ray. Berkeley: University of California Press.

Hurwitz, Jon and Mark Peffley. 2005. "Playing the Race Card in the Post-Willie Horton Era: The Impact of Racialized Code Words on Support for Punitive Crime Policy." *Public Opinion Quarterly* 69(1):99–112.

Ingram, Christopher. 2014. "White People Are More Likely to Deal Drugs, but Black People Are More Likely to Get Arrested for It". *Washington Post*. Accessed: https://www.washingtonpost.com/news/wonk/wp/2014/09/30/white-people-are-more-likely-to-deal-drugs-but-black-people-are-more-likely-to-get-arrested-for-it.

Ingram, Christopher. 2017. "Public Support for Marijuana Legalization Surged in 2016". *Washington Post*. Accessed: https://www.washingtonpost.com/news/wonk/wp/2017/03/29/public-support-for-marijuana-legalization-surged-in-2016.

Jackson, Andrew. 1833. "Fifth Annual Message", Dec 3. Online by Gerhard Peters and John T. Woolley, *The American Presidency Project*. http://www.presidency.ucsb.edu/ws/?pid=29475.

Jackson, Michael. 2013. "The Myth of the Black-on-Black Crime Epidemic". *Demos*. Accessed: http://www.demos.org/blog/7/29/13/myth-black-black-crime-epidemic.

Jacobs, Ronald N. and Eleanor Townsley. 2011. *The Space of Opinion: Media Intellectuals and the Public Sphere*. New York, NY: Oxford University Press.

Jacobson, Matthew Frye. 1999. *Whiteness of a Different Color: European Immigrants and the Alchemy of Race*. Cambridge, MA: Harvard University Press.

Johnson, Carrie. 2014. "20 Years Later, Parts of Major Crime Bill Viewed as Terrible Mistake". *NPR* Sep 12. Accessed: http://www.npr.org/2014/09/12/347736999/20-years-later-major-crime-bill-viewed-as-terrible-mistake.

Jordan, Winthrop D. 1968. *White over Black: American Attitudes toward the Negro, 1550–1812*. Chapel Hill: University of North Carolina Press.

Kapell, Matthew Wilhelm. 2009. "'Miscreants be They White or Colored': The Local Press Reactions to the 1943 Detroit 'Race Riot'". *Michigan Academician* XXXIX:213–244.

Kelley, Robin D. G. 1997. *Yo' Mama's Disfunktional!: Fighting the Culture Wars in Urban America*. Boston, MA: Beacon Press.
King, Mike. 2017. "Aggrieved Whiteness: White Identity Politics and Modern American Racial Formation". *Abolition*. Accessed: https://abolitionjournal.org/aggrieved-whiteness-white-identity-politics-and-modern-american-racial-formation/.
Kraus, Michael W. and Bennett Callaghan. 2016. "Social Class and Prosocial Behavior: The Moderating Role of Public versus Private Contexts". *Social Psychological and Personality Science* 7(8):769–777.
Kraus, Michael W., Stéphane Côté, and Dacher Keltner. 2010. "Social Class, Contextualism, and Empathic Accuracy". *Psychological Science* 21(11):1716–1723.
La Porte, Amy. 2016. "Spike in Hate Crimes Prompts Special NY Police Unit." CNN. Accessed: http://www.cnn.com/2016/11/20/us/hate-crime-unit-new-york/index.html.
Lagisetty, Pooja A., Ryan Ross, Amy Bohnert, Michael Clay, and Donovan T. Maust. 2019. "Buprenorphine Treatment Divide by Race/Ethnicity and Payment." *JAMA Psychiatry* 76(9):979–981.
Lamont, Michèle. 2000. *The Dignity of Working Men: Morality and the Boundaries of Race, Class, and Immigration*. Cambridge, MA: Harvard University Press.
Lamont, Michéle and Virág Molnár. 2002. "The Study of Boundaries in the Social Sciences". *Annual Review of Sociology* 28:167–195.
Lareau, Annette. 2011. *Unequal Childhoods: Class, Race, and Family Life*. 2nd Edition. Berkeley: University of California Press.
Lears, T. J. Jackson. 1985. "The Concept of Cultural Hegemony: Problems and Possibilities". *The American Historical Review* 90(3):67–593.
Lerner, Melvin. 1980. *Belief in a Just World: A Fundamental Delusion*. Berlin, Germany: Springer.
Levy-Pounds, Nekima. 2010. "Can These Bones Live? A Look at the Impacts of the War on Drugs on Poor African-American Children and Families". *Hastings Race and Poverty Law Journal* 7(2):353–380.
Lewis, Amanda E. 2003a. "Everyday Race Making: Navigating Racial Boundaries in Schools". *American Behavioral Scientist* 47(3):283–305.
Lewis, Amanda E. 2003b. *Race in the Schoolyard: Negotiating the Color Line in Classrooms and Communities*. New Brunswick, NJ: Rutgers University Press.
Lewis, Amanda E. 2004. ""What Group?" Studying Whites and Whiteness in the Era of "Color-Blindness"". *Sociological Theory* 22(4):623–646.
Lewis, Amanda Chicago. 2016. "How Black People Are Being Shut Out of America's Weed Boom: Whitewashing the Green Rush". *Buzzfeed* Mar 26. Accessed: https://www.buzzfeed.com/amandachicagolewis/americas-white-only-weed-boom.

Lewis, Oscar. 1966. "The Culture of Poverty". *Scientific American* 215(4):19–25.
Lopez, German. 2019. "What Alex Berenson's New Book Gets Wrong about Marijuana, Psychosis, and Violence." *Vox* Jan 14. Accessed: https://www.vox.com/future-perfect/2019/1/14/18175446/alex-berenson-tell-your-children-marijuana-psychosis-violence.
Los Angeles Herald. 1875a. "News of the Morning". *California Digital Newspaper Collection* Nov 19. Accessed: https://cdnc.ucr.edu/?a=d&d=LAH18751119.2.3&srpos=1&e=------187-en--20-LAH-1--txt-txIN-Opium+dens----1875---1.
Los Angeles Herald. 1875b. "Latest Telegrams". *California Digital Newspaper Collection* Dec 7. Accessed: https://cdnc.ucr.edu/?a=d&d=LAH18751207.2.6&srpos=3&e=------187-en--20-LAH-1--txt-txIN-Opium+dens----1875---1.
Lucas, Ryan. 2017. "Trump Administration Lifts Limits on Military Hardware for Police". NPR, Aug 28. Located: https://www.npr.org/2017/08/28/546743742/trump-administration-lifts-limits-on-military-hardware-for-police
Luke, Carmen. 1994. "Feminist Pedagogy and Critical Media Literacy". *Journal of Communication Inquiry* 18(2):30–47.
Lurigio, Arthur J. and Pamela Loose. 2008. "The Disproportionate Incarceration of African Americans for Drug Offenses: The National and Illinois Perspective". *Journal of Ethnicity and Criminal Justice* 3:223–247.
Lutz, Ashley. 2012. "These 6 Corporations Control 90% of the Media in America." *Business Insider* Jun 14. Accessed: https://www.businessinsider.com/these-6-corporations-control-90-of-the-media-in-america-2012-6.
Lynch, Mona, Marisa Omori, Aaron Roussell, and Matthew Valasik. 2013. "Policing the 'Progressive' City: The Racialized Geography of Drug Law Enforcement". *Theoretical Criminology* 17(3):335–357.
MacLear, Michael. 1982. *Ten Thousand Day War: Vietnam: 1945–1975*. New York, NY: Avon Books.
Maldovan, Theodora. 2016. "The Six Companies That Own (Almost) All Media". *Affinity Magazine*. Accessed: http://affinitymagazine.us/2016/12/07/the-six-companies-that-own-almost-all-media/.
Males, Mike. 2013. "Why the Gigantic, Decades-Long Drop in Black Youth Crime Threatens Major Interests." *Center on Juvenile and Criminal Justice*. Accessed: http://www.cjcj.org/news/6523.
Mann, Brian. 2013. "Profile: Charles Rangel and the Drug Wars". *WNYC*. Accessed: http://www.wnyc.org/story/313060-profile-charles-rangel-and-drug-wars/.
Marni, Davis. 2012. *Jews and Booze: Becoming American in the Age of Prohibition*. New York, NY: New York University Press.

Marshall, Amani. 2010. "'They Will Endeavor to Pass for Free': Enslaved Runaways' Performances of Freedom in Antebellum South Carolina". *Slavery & Abolition* 31(2):161–180.

Martín Sánchez-Jankowski. 2008. *Cracks in the Pavement: Social Change and Resilience in Poor Neighborhoods.* Berkeley: University of California Press.

Martin, Gregory J. and Joshua McCrain. 2019. "Local News and National Politics." *American Political Science Review* 113(2):372–384.

Martinez, Jr., Ramiro and Matthew T. Lee. 2000. "On Immigration and Crime". Pp. 485–524 in *Criminal Justice 2000, Vol. 1: The Nature of Crime: Continuity and Change.* Edited by Gary LaFree. Washington, DC: National Institute of Justice.

Marx, Karl and Friedrich Engels. 1978[1845–1846]. "The German Ideology". Pp.146–200 in The *Marx-Engels Reader.* Edited by Robert C. Tucker. New York, NY: W. W. Norton & Company.

Massey, Douglass and Nancy Denton. 1993. *American Apartheid: Segregation and the Making of the Underclass.* Cambridge, MA: Harvard University Press.

Mastro, Dana E. and Linda R. Tropp. 2004. "The Effects of Interracial Contacts, Attitudes, and Stereotypical Portrayals on Evaluations of Black Sitcom Characters." *Communication Research Reports* 21(2):119–129.

Matheson, Donald. 2005. *Media Discourses: Analysing Media Texts.* London: McGraw-Hill Education.

Mathias, Christopher. 2014. "Ramarley Graham was Killed by The NYPD over Two Years ago, and Nothing Has Happened". *The Huffington Post* Aug 8, 2014. Accessed: http://www.huffingtonpost.com/2014/08/08/ramarley-graham-nypd_n_5662134.html.

Matthews, Dylan. 2015. "Woodrow Wilson was Extremely Racist – Even by the Standards of His Time". *Vox.* Accessed: https://www.vox.com/policy-and-politics/2015/11/20/9766896/woodrow-wilson-racist.

Mayor's Committee on Marihuana. 1944. *The Marihuana Problem in the City of New York; Sociological, Medical, Psychological and Pharmacological Studies.* Lancaster, PA: Jacques Cattell.

McCombs, Maxwell, Donald L. Shaw, and David Weaver. 1997. *Communication and Democracy: Exploring Intellectual Frontiers in Agenda-Setting Theory.* New York, NY: Routledge.

McCormick, Anita Louise. 2000. *The Vietnam Antiwar Movement in American History.* Berkeley Heights, NJ: Enslow Publishing.

Mead, George Herbert. 1934. *Mind, Self, and Society.* Chicago, IL: University of Chicago Press.

Meads, Mallory. 2016. "The War against Ourselves: Heien v. North Carolina, the War on Drugs, and Police Militarization." *University of Miami Law Review* 70:615–647.

Mendelberg, Tali. 2001. *The Race Card: Campaign Strategy, Implicit Messages, and the Norm of Equality*. Princeton, NJ: Princeton University Press.
Merriam Webster. 2016. "Racism Definition". Located online: https://www.merriam-webster.com/dictionary/racism
Merton, Robert K. 1932. "Social Structure and Anomie". *American Sociological Review* 3(5):672–682.
Merton, Robert K. 1949. *Social Theory and Social Structure*. New York, NY: Simon and Schuster.
Meyers, Kristen A. 2005. *Racetalk: Racism Hiding in Plain Sight*. Lanham, MD: Rowman & Littlefield Publishers.
Miami Herald. 2016. "'Code' Lets Social Media Users Hide Racist Slurs". *Miami Herald* Oct 1. Accessed: http://www.miamiherald.com/news/nation-world/national/article105476826.html.
Mitchell, Ojmarrh and Michael S. Caudy. 2013. "Examining Racial Disparities in Drug Arrests". *Justice Quarterly* 32(2): 288–313.
Mill, John Stuart. 1859. *On Liberty*. London: J.W. Parker.
Mills, Charles W. 2007. "White Ignorance." Pp. 11–38 in *Race and Epistemologies of Ignorance*. Edited by Shannon Sullivan and Nancy Tuana. Albany: State University of New York Press.
Mills, C. Wright. 1959. *The Sociological Imagination*. New York, NY: Oxford University Press.
Morgan, Wayne H. 1981. *Drugs in America: A Social History, 1800–1980*. Syracuse, NY: Syracuse University Press.
Morris, Aldon D. 1984. *The Origins of the Civil Rights Movement*. NY: Free Press.
Morris, Aldon. 2015. *The Scholar Denied: W.E.B. Du Bois and the Birth of Modern Sociology*. Oakland: University of California Press.
Moynihan, Daniel Patrick. 1965. *The Negro Family: A Case for National Action*. Washington, DC: U.S. Department of Labor.
Mueller, Jennifer C. 2017. "Producing Colorblindness: Everyday Mechanisms of White Ignorance". *Social Problems* 64(2):219–238.
Muhammad, Khalil Gibran. 2011. *The Condemnation of Blackness: Race, Crime, and the Making of Modern Urban America*. Cambridge, MA: Harvard University Press.
Murphey, Lynne. 2016. "Linguistics Explains Why Trump Sounds Racist When He Says "The" African Americans". *Quartz* Oct 11. Accessed: http://qz.com/806174/second-presidential-debate-linguistics-explains-why-donald-trump-sounds-racist-when-he-says-the-african-americans/.
Netherland, Julie and Helena B. Hansen. 2016. "The War on Drugs That Wasn't: Wasted Whiteness, "Dirty Doctors," and Race in Media Coverage of Prescription Opioid Misuse". *Culture, Medicine, and Psychiatry* 40(23):664–686. DOI: 10.1007/s11013-016-9496-5.

Neubeck, Kenneth J. and Noel A. Cazenave. 2001. *Welfare Racism: Playing the Race Card Against America's Poor.* New York, NY: Routledge.

Nicosia, Nancy, John MacDonald, and Jeremy Arkes. 2013. "Disparities in Criminal Court Referrals to Drug Treatment and Prison for Minority Men". *American Journal of Public Health* 103(6):e77–e84.

Nietzsche, Friedrich. 1886 [1907]. *Beyond Good and Evil: Prelude to a Philosophy of the Future.* Translated by Helen Zimmern. New York, NY: MacMillan Company.

Nixon, Richard. 1971. "Special Message to the Congress on Drug Abuse Prevention and Control". The American Presidency Project Jun 17. Online by Gerhard Peters and John T. Woolley. http://www.presidency.ucsb.edu/ws/?pid=3048.

Norton, Michael I. and Samuel R. Sommers. 2011. "Whites See Racism as a Zero-Sum Game That They are Now Losing." *Perspectives on Psychological Science* 6(3):215–218.

NPR. 2016. "As Adults Legally Smoke Pot in Colorado, More Minority Kids Arrested for It". *npr.org* Jun 26. Accessed: http://www.npr.org/2016/06/29/483954157/as-adults-legally-smoke-pot-in-colorado-more-minority-kids-arrested-for-it.

NYCLU. 2016. "Stop-and-Frisk Data". *New York Civil Liberties Union.* Accessed: http://www.nyclu.org/content/stop-and-frisk-data.

Obasogie, Osagie. 2013. *Blinded by Sight: Seeing Race through the Eyes of the Blind.* Stanford, CA: Stanford University Press.

Office of the Attorney General. 2017. *Memorandum for All Federal Prosecutors.* Washington, DC: U.S. Department of Justice.

Ojmarrh, Mitchell and Michael S. Caudy. 2013. "Examining Racial Disparities in Drug Arrests". *Justice Quarterly* 32(2):288–313.

Okeowo, Alexis. 2016. "Hate on the Rise after Trump's Election". *The New Yorker.* Accessed: http://www.newyorker.com/news/news-desk/hate-on-the-rise-after-trumps-election.

Oliver, Melvin L. and Thomas M. Shapiro. 1995. *Black Wealth / White Wealth: A New Perspective on Racial Inequality.* New York, NY: Routledge.

Oliver, Melvin L. and Thomas M. Shapiro. 2019. "Disrupting the Racial Wealth Gap". *Contexts* 18(1):16–21.

Omi, Michael and Howard Winant. 2014. *Racial Formation in the United States.* 3rd Edition. New York, NY: Routledge.

Osuma, Steven. 2020. "Transnational Moral Panic: Neoliberalism and the Spectre of MS-13". *Race & Class.* DOI: 10.1177/0306396820904304.

Pager, Devah. 2007. *Marked: Race, Crime, and Finding Work in an Era of Mass Incarceration.* Chicago, IL: University of Chicago Press.

Paley, Dawn. 2015. "Drug War as Neoliberal Trojan Horse". *Latin American Perspectives.* DOI: 10.1177/0094582X15585117.

Parlett, Martin A. 2014. *Demonizing a President: The "Foreignization" of Barrack Obama*. Denver, CO: Praeger.

Parsons, Talcott. 1961. "An Outline of the Social System". Pp. 30–84 in *Theories of Society: Foundations of Modern Sociological Theory, Volume II*. Edited by Talcott Parsons, Edward Shils, Kaspar D. Naegele, and Jesse R. Pitts. New York, NY: The Free Press.

Patterson, Orlando. 1982. *Slavery and Social Death: A Comparative Study*. Cambridge, MA: Harvard University Press.

Pew Research Center. 2014. "America's New Drug Policy Landscape". *Pew Research Center* Apr 2. Accessed: http://www.people-press.org/2014/04/02/americas-new-drug-policy-landscape/.

Pew Research Center. 2016a. "Social Media Conversations about Race". *Pew Research Center Report* Aug 15. Accessed: http://www.pewinternet.org/2016/08/15/social-media-conversations-about-race/.

Pew Research Center. 2016b. "Support for Marijuana Legalization Continues to Rise". *Pew Research Center* Oct 12. Accessed: http://www.pewresearch.org/fact-tank/2016/10/12/support-for-marijuana-legalization-continues-to-rise/.

Pfohl, Stephen J. 1985. *Images of Deviance & Social Control: A Sociological History*. New York, NY: McGraw Hill.

Phelan, Jo C., Bruce G. Link, and Naumi M. Feldman. 2013. "The Genomic Revolution and Beliefs about Essential Racial Differences: A Backdoor to Eugenics?". *American Sociological Review* 78(2):167–191.

Phelps, Jordyn. 2016. "Maine Gov. LePage Says Drug Dealers Have Names Like 'D-Money' and 'Impregnate...White Girl[s]'". *ABCNews* Jan 7. Accessed: abcnews.go.com/Politics/maine-gov-lepage-drug-dealers-names-money-impregnatewhite/.

Picca, Leslie Houts and Joe R. Feagin. 2007. *Two-Faced Racism: Whites in the Backstage and Frontstage*. New York, NY: Routledge.

Piff, Paul K., Michael W. Kraus, Stéphane Côté, Bonnie Hayden Cheng, and Dacher Keltner. 2010. "Having Less, Giving More: The Influence of Social Class on Prosocial Behavior". *Journal of Personality and Social Psychology* 99(5):771–784.

Piff, Paul K., Daniel M. Stancato, Stéphane Côté, Rodolfo Mendoza-Denton, and Dacher Keltner. 2012. "Higher Social Class Predicts Increased Unethical Behavior". *Proceedings of the National Academy of Science* 109(11):4086–4091.

Pride, Armistead S. and Clint C. Wilson II. 1997. *A History of the Black Press*. Washington, DC: Howard University Press.

Provine, Marie Doris. 2007. *Unequal under Law: Race in the War on Drugs*. Chicago, IL: University of Chicago Press.

Rabaka, Reiland. 2010. *Against Epistemic Apartheid: W.E.B. Du Bois and the Disciplinary Decadence of Sociology*. Lanham, MD: Lexington Books.

RabbitEars. 2017. "Stations for Owner – Sinclair". Accessed: http://rabbitears.info/search.php?request=owner_search&owner=Sinclair&sort=state.

Reilly, Kate. 2016. "Racist Incidents are up since Donald Trump's Election. These are Just a Few of Them." *Time*. Accessed: http://time.com/4569129/racist-anti-semitic-incidents-donald-trump/.

Reiman, Jeffery. 2001. *The Rich Get Richer and the Poor Get Prison: Ideology, Class, and Criminal Justice*. 6th Edition. Boston, MA: Allyn and Bacon.

Ridgeway, Cecilia L. and Tamar Kricheli-Katz. 2013. "Intersecting Cultural Beliefs in Social Relations: Gender, Race, and Class Binds and Freedoms". *Gender & Society* 27(3):294–318.

Rios, Victor. 2011. *Punished: Policing the Lives of Young Black and Latino Boys*. New York, NY: New York University Press.

Roberts, Bryan R. and Yu Chen. 2013. "Drugs, Violence, and the State". *Annual Review of Sociology* 39:105–125.

Roberts, Dorothy. 2011. *Fatal Invention: How Science, Politics, and Big Business Re-Create Race in the Twenty-First Century*. New York, NY: The New Press.

Robins, Kevin. 1997. "What in the World's Going on?". Pp. 12–45 in *Production of Culture/Cultures of Production*. Edited by Paul du Gay. Thousand Oaks, CA: Sage.

Robinson, Laura, Shelia R. Cotton, Hiroshi Ono, Anabel Quan-Haase, Gustavo Mesch, Wenhong Chen, Jeremy Schulz, Timothy M. Hale, and Michael J. Stern. 2015. "Digital Inequalities and Why They Matter". *Information, Communication & Society* 18(5):569–582.

Robinson, Matthew. 2000. "The Construction and Reinforcement of Myths of Race and Crime". *Journal of Contemporary Criminal Justice* 16(2):133–156.

Roeder, Oliver, Lauren-Brooke Eisen, and Julia Bowling. 2015. *What Caused the Crime Decline?* New York: Brennan Center for Justice at New York University School of Law.

Roediger, David R. 1991. *The Wages of Whiteness: Race and the Making of the American Working Class*. London: Verso.

Rogers, Everett M. and James W. Dearing. 1988. "Agenda-Setting Research: Where Has It Been, Where Is It Going?". *Communication Yearbook* 11:555–594.

Rogers, Everett M., William B. Hart, and James W. Dearing. 1997. "A Paradigmatic History of Agenda-Setting Research". Pp. 225–236 in *Do the Media Govern?: Politicians, Voters, and Reporters in America*. Edited by Shanto Iyengar and Richard Reeves. New York, NY: Sage.

Romo, Vanessa. 2019. "As Illinois Prepares to Legalize Pot, Public Housing Tenants Not Allowed to Partake". *NPR* Nov 11. Accessed: https://www.npr.org/2019/11/11/778371751/as-illinois-prepares-to-legalize-pot-public-housing-tenants-not-allowed-to-parta.

Rose, Dina R., and Todd R. Clear. 1998. "Incarceration, Social Capital and Crime: Implications for Social Disorganization Theory". *Criminology* 36(3):441–480.

Rosen, Jay. 2006. "The People Formerly Known as the Audience." *Pressthink* Jun 27. Accessed: http://archive.pressthink.org/2006/06/27/ppl_frmr.html.

Rosino, Michael L. 2016. "Boundaries and Barriers: Racialized Dynamics of Political Power". *Sociology Compass* 10(10):939–951.

Rosino, Michael L. 2017. "Dramaturgical Domination: The Genesis and Evolution of the Racialized Interaction Order". *Humanity & Society* 41(2):158–181.

Rosino, Michael L. and Matthew W. Hughey. 2016. ""Who's Invited to the (Political) Party: Race and Party Politics in the USA." *Ethnic and Racial Studies* 39(3):325–332.

Rosino, Michael L. and Matthew W. Hughey. 2017. "Speaking through Silence: Racial Discourse and Identity Construction in Mass Mediated Debates on the 'War on Drugs'". *Social Currents* 4(3):246–264.

Ross, Karen and Virginia Nightingale. 2003. *Media and Audiences: New Perspectives*. Buckingham, UK: Open University Press.

Rumbaut, Reubén G. 2009. "Undocumented Immigration and Rates of Crime and Imprisonment: Popular Myths and Empirical Realities". Pp. 119–139 in *The Role of Local Police: Striking a Balance between Immigration Enforcement and Civil Liberties*. Edited by Anita Khashu. Washington, DC: Police Foundation.

Ryan, William. 1976. *Blaming the Victim*. Revised Edition. New York, NY: Vintage.

Sandy, Kathleen R. 2003. "The Discrimination Inherent in America's Drug War: Hidden Racism Revealed by Examining the Hysteria over Crack". *Alabama Law Review* 54:665–693.

San Francisco Office of Cannabis. 2019. "Equity Program". Located online at: https://officeofcannabis.sfgov.org/equity

Schlesinger, Traci. 2013. "Racial Disparities in Pretrial Diversion: An Analysis of Outcomes among Men Charged with Felonies and Processed in State Courts". *Race and Justice* 3(3):210–238.

Schudson, Michael. 1978. *Discovering the News: A Social History of American Newspapers*. New York, NY: Basic Books.

Schudson, Michael. 1989. "How Culture Works: Perspectives from Media Studies on the Efficacy of Symbols." *Theory and Society* 18(2):153–180.

Schudson, Michael. 2011. *The Sociology of News*. 2nd Edition. New York, NY: W.W. Norton Co.

Schultz, Kevin M. 2015. *Buckley and Mailer: The Difficult Friendship That Shaped the Sixties*. New York, NY: W.W. Norton and Company.

Schwalbe, Michael. 2005. *The Sociologically Examined Life: Pieces of the Conversation*. New York, NY: McGraw-Hill.

Schwalbe, Michael. 2008. *Rigging the Game: How Inequality Is Reproduced in Everyday Life*. New York, NY: Oxford University Press.

Seamster, Louise. 2019. "Black Debt, White Debt." *Contexts* 18(1):30–35.

Seibert, Darcy Clay, Dina J. Wilke, Jorge Delva, Michael P. Smith, and Richard L. Howell. 2003. "Differences in African American and White College Students' Drinking Behaviors: Consequences, Harm Reduction Strategies, and Health Information Sources". *Journal of American College Health* 52(3):123–129.

Sewell, Abigail A. and Kevin A. Jefferson. 2016. "Collateral Damage: The Health Effects of Invasive Police Encounters in New York City". *Journal of Urban Health* 93(1):42–67.

Shaw, Clifford R. and Henry D. McKay. 1942. *Juvenile Delinquency in Urban Areas*. Chicago, IL: University of Chicago Press.

Silverstein, Joel. 2013. "I Don't Feel Your Pain: A Failure of Empathy Perpetuates Racial Disparities". *Slate* Jun 27. Accessed: http://www.slate.com/articles/health_and_science/science/2013/06/racial_empathy_gap_people_don_t_perceive_pain_in_other_races.html.

Sitkoff, Harvard. 1969. "The Detroit Race Riots of 1943." *Michigan History* 53:183–206.

Smetko, Holli A. and Patti M. Valkenburg. 2000. "Framing European Politics: A Content Analysis of Press and Television News". *Journal of Communication* 50(2):93–109.

Solórzano, Daniel G. and Tara J. Yosso. 2002. "Critical Race Methodology: Counter-Storytelling as an Analytical Framework for Education Research". *Qualitative Inquiry* 8(2):23–44.

Sonnad, Nikhil. 2016. "Alt-Right Trolls Are Using These Code Words for Racial Slurs Online". *Quartz* Oct 1. Accessed: http://qz.com/798305/alt-right-trolls-are-using-googles-yahoos-skittles-and-skypes-as-code-words-for-racial-slurs-on-twitter/.

Southern Poverty Law Center. 2017. "Ten Days after: Harassment and Intimidation in the Aftermath of the Election". Accessed: https://www.splcenter.org/20161129/ten-days-after-harassment-and-intimidation-aftermath-election.

Stack, Carol B. 1974. *All Our Kin: Strategies for Survival in a Black Community*. New York, NY: Harper-Collins.

Statista. 2017. "Average Time Spent with Major Media per Day in the United States as of April 2017 (in minutes)". Accessed: https://www.statista.com/statistics/276683/media-use-in-the-us/.

Steinberg, Stephen. 1989. *The Ethnic Myth: Race, Ethnicity, and Class in America*. 2nd Edition. Boston, MA: Beacon Press.

Steinberg, Stephen. 1995. *Turning Back: The Retreat from Racial Justice in American thought and Policy*. Boston, MA: Beacon Press.

Steinberg, Stephen. 2007. *Race Relations: A Critique*. Stanford, CA: Stanford University Press.

Steinberg, Stephen. 2011. "Poor Reason: Culture Still Doesn't Explain Poverty". *Boston Review*. Accessed: http://bostonreview.net/steinberg.php.

Steiner, Tobias. 2015. "Under the Macroscope: Convergence in the US Television Market Between 2000 and 2014". *Image* 22(3):4–21.

Stoughton, Seth. 2015. "Law Enforcement's "Warrior" Problem". *Harvard Law Review Forum* 128(255):225–264.

Sugrue, Thomas J. 2005. *The Origins of Urban Crisis: Race and Inequality in Postwar Detroit*. Princeton, NJ: Princeton University Press.

Sullum, Jacob. 2004. *Saying Yes: In Defense of Drug Use*. New York, NY: TarcherPerigee.

Szoldra, Paul. 2014. "This is the Terrifying Result of the Militarization of Police". *Business Insider* Aug 12. Accessed: http://www.businessinsider.com/police-militarization-ferguson-2014-8.

Taslitz, Andrew E. 2013. "Racial Threat versus Racial Empathy in Sentencing-Capital and Otherwise". *American Journal of Criminal Law* 41(1):1–40.

Taylor, Carol M. 1981. "W.E.B. DuBois's Challenge to Scientific Racism". *Journal of Black Studies* 11(4):449–460.

Tonry, Michael. 1994. "Race and the War on Drugs". *University of Chicago Legal Forum* 1994:25–82.

The National Commission on Marihuana and Drug Use. 1972. *Marihuana: A Signal of Misunderstanding, First Report of the National Commission on Marihuana and Drug Use*. Washington, DC: U.S. Government Printing Office.

The Onion. 1998. "Drugs Win Drug War". *The Onion* Jan 10, p. 1. Accessed: https://legacy.npr.org/assets/news/2013/onion-drugs.pdf?t=1580465816463

The Sentencing Project. 2014. *Race and Punishment: Racial Perceptions of Crime and Support for Punitive Policies*. Washington, DC: The Sentencing Project.

Thomas, James and David Brunsma. 2014. "Oh, You're Racist? I've Got a Cure for That!" *Ethnic and Racial Studies* 37(9):1467–1485.

Trujillo, Josmar. 2012. "Media Laugh Off Criticisms of Drug War". *Extra!* Dec 2012:6–7.

Ture, Kwame and Charles V. Hamilton. 1967 [1992]. *Black Power: The Politics of Liberation*. New York, NY: Vintage Books.

Turner, S. Derek. 2007. "Out of the Picture: The Lack of Racial and Gender Diversity in TV Station Ownership". Pp. 181–194 in *The Case against Media Consolidation: Evidence on Concentration, Localism and Diversity*. Edited by Mark N. Cooper. New York, NY: Donald McGannon Center for Communications Research, Fordham University.

Turner, S. Derek and Mark N. Cooper. 2007. "The Lack of Racial and Gender Diversity in Broadcast Ownership & the Effects of FCC Policy:

An Empirical Analysis." Paper presented at TRPC 2007. Accessed: http://papers.ssrn.com/sol3/papers.cfm?abstract_id=2113819.

Twine, France Winddance. 2004. "A White Side of Black Britain: The Concept of Racial Literacy". *Ethnic and Racial Studies* 27(6): 878–907.

Twine, France Winddance. 2010. *A White Side of Black Britain: Interracial Intimacy and Racial Literacy.* Durham, NC: Duke University Press.

United States Committee on the Judiciary. 1974. *Marihuana-Hashish Epidemic and Its Impact on United States Security, Hearings before the Subcommittee to Investigate the Administration of the Internal Security Act and other Internal Security Laws of the Committee on the Judiciary, 93rd Congress Second Session.* Washington, DC: U.S. Government Printing Office.

United States Congress. 1944. "Congressional Record – Senate, June 20th". Pp. 6259 in *Congressional Record: Proceedings and Debates of the 78th Congress Second Session.* Washington, DC: U.S. Government Printing Office.

United States Department of Justice. 2015. "In Milestone for Sentencing Reform, Attorney General Holder Announces Record Reduction in Mandatory Minimums against Nonviolent Drug Offenders". Memo released Tuesday, February 17, 2015. Accessed: https://www.justice.gov/opa/pr/milestone-sentencing-reform-attorney-general-holder-announces-record-reduction-mandatory.

United States Department of Justice. 2017. "Attorney General Jeff Sessions Delivers Remarks at the 30th DARE Training Conference". Remarks as prepared for delivery on Tuesday, July 11, 2017. Accessed: https://www.justice.gov/opa/speech/attorney-general-jeff-sessions-delivers-remarks-30th-dare-training-conference.

VanDerWerff, Emily. 2019. "Here's What Disney Owns after the Massive Disney/Fox Merger". *Vox* Mar 20. Accessed: https://www.vox.com/culture/2019/3/20/18273477/disney-fox-\merger-deal-details-marvel-x-men.

van Dijk, Teun A. 1992. "Discourse and the Denial of Racism". *Discourse & Society* 3(1):87–118.

van Dijk, Teun A. 2008. *Discourse and Power.* Basingstoke, UK: Pelgrave MacMillan.

Vergne, Derick E. 2016. "Apathy, Explained". *MedScape* Mar 16. Accessed: http://www.medscape.com/viewarticle/860177_5.

Vizcarrondo, Tom. 2013. "Measuring Concentration of Media Ownership: 1976–2009". *International Journal on Media Management* 15(3):177–195.

Voigt, Rob, Nicholas P. Camp, Vinodkumar Prabhakaran, William L. Hamilton, Rebecca C. Hetey, Camilla M. Griffiths, David Jurgen, Dan Jurafsky and Jennifer L. Eberhardt. 2017. "Language from Police Body

Camera Footage Shows Racial Disparities in Officer Respect". *Proceedings of the National Academy of the Sciences* 114(25):6521–6526.
Vuolo, Mike, Joy Kadowaki, and Brian C. Kelly. 2017. "Marijuana's Moral Entrepreneurs, Then and Now". *Contexts* Sep 14. Accessed: https://contexts.org/articles/marijuanas-moral-entrepreneurs/.
Wacquant, Loïc. 2001. "Deadly Symbiosis: When Ghetto and Prison Meet and Mesh". *Punishment and Society* 3(1):95–134.
Wacquant, Loïc. 2009. *Punishing the Poor: The Neoliberal Government of Social Insecurity*. Durham, NC: Duke University Press.
Wacquant, Loïc. 2010. "Class, Race, & Hyperincarceration in Revanchist America". *Daedalus* 193(3):74–146.
Walsh, Katherine Cramer. 2004. *Talking about Politics: Informal Groups and Social Identity in American Life*. Chicago, IL: University of Chicago Press.
Ward, Jeffrey T., Richard D. Hartley, and Rob Tillyer. 2016. "Unpacking Gender and Racial/Ethnic Biases in the Federal Sentencing of Drug Offenders: A Causal Mediation Approach". *Journal of Criminal Justice* 46:196–206.
Washington, George. 1783. "Letter to James Duane". *National Archives*. Washington, DC: U.S. Library of Congress. Accessed: https://founders.archives.gov/documents/Washington/99-01-02-11798
Webb, Gary. 1999. *Dark Alliance: The CIA, the Contras, and the Cocaine Explosion*. 2nd Edition. New York, NY: Seven Stories Press.
Weber, Max. 1958. *The Protestant Ethic and the Spirit of Capitalism*. Translated by Talcott Parsons. New York, NY: Charles Scribner's Sons.
Weedon, Chris. 1989. *Feminist Practice and Poststructuralist Theory*. London: Basil Blackwell.
Weekly Butte Record. 1877. "Opium Smoking". *California Digital Newspaper Collection* Mar 3. Accessed: https://cdnc.ucr.edu/?a=d&d=WBR18770303.2.40&srpos=1&e=------187-en--20--1--txt-txIN-Opium+dens-ARTICLE------1.
Welch, Kelly. 2007. "Black Criminal Stereotypes and Racial Profiling". *Journal of Contemporary Criminal Justice* 23:276–288.
Wellstone, Paul. 2000. "Growing Media Consolidation Must Be Examined to Preserve Our Democracy". *Federal Communications Law Journal* 52:551–554.
Western, Bruce. 2007. *Punishment and Inequality in America*. New York, NY: Russel Sage.
White, Davon. 2017. "Ramarley Graham (1993–2012)". *BlackPast.org* Nov 10. Accessed: https://www.blackpast.org/african-american-history/graham-ramarley-1993-2012/.
Williams, Janice. 2017. "Jeff Sessions on Marijuana: Drug is 'Only Slightly Less Awful' Than Heroin". *Newsweek* Mar 15. Accessed: http://www.newsweek.com/jeff-sessions-marijuana-legalization-states-heroin-opioids-568499.

WNYC. 2016. "Ramarley Graham's Mom: 'Their Lives Don't Matter'". *WNYC News*. Accessed: http://www.wnyc.org/story/feds-will-not-prosecute-nypd-officers-shooting-death-ramarley-graham/.

Wright II, Earl. 2002. "Using the Master's Tools: The Atlanta Sociological Laboratory and American Sociology, 1896–1924". *Sociological Spectrum* 22:15–39.

Yang, Tim. 2004. "The Malleable Yet Undying Nature of the Yellow Peril". Accessed: http://www.dartmouth.edu/~hist32/History/S22%20-The%20Malleable%20Yet%20Undying%20Nature%20of%20the%20Yellow%20Peril.htM.

Young, Robert. 1971. "Nixon Declares War on Narcotics Use in the U.S." *Chicago Tribune* Jun 18. Accessed: https://chicagotribune.newspapers.com/image/201487138/?terms=Nixon%2BDeclares%2BWar

Zernike, Kate. 2011. "The Persistence of Conspiracy Theories". *The New York Times*. Accessed: http://www.nytimes.com/2011/05/01/weekinreview/01conspiracy.html.

Zerubavel, Eviatar. 2006. *The Elephant in the Room: Silence and Denial in Everyday Life*. New York, NY: Oxford University Press.

Glossary

1981 Military Cooperation with Law Enforcement Act An Act passed under the Regan Administration that created more possibilities for domestic military operations, allowing military weaponry, equipment, training, and intelligence to pour into police departments aiding in the militarization of drug policy enforcement.

1994 Violent Crime Control and Law Enforcement Act An Act passed under the Clinton Administration that exasperated the rate of incarceration in the United States and further racialized the prison population through provisions such as the "Three Strikes Law."

Affect The experience of emotions, sentiments, and feelings in everyday social life (see Thomas and Brunsma 2013).

Agency The ability to make choices given the range of possibilities as determined by the social structure (see entry for **Structure**; Giddens 1993).

Agenda-Setting Agenda-setting occurs when people use their power or influence to determine what particular events, issues, or problems should be discussed, debated, and focused on in society.

Audiences The groups of people that witness, consume, and even participate in media events and increasingly communicate with media content and each other as technology advances (see Ross and Nightingale 2003)

Code Words Words, concepts, and phrases that suggest a particular racial group without explicitly naming that group or category (see entry for **Group-Based Frames**; Hurwitz and Peffley 2005; Haney Lopez 2014; Hughey and Parks 2014).

Collective Action Efforts by groups in society to act together, often to change a norm, rule, or social arrangement, and often in ways that are organized.

Colorblind Defense Frame A way of rationalizing the War on Drugs without overt reference to racial categories (see entry for **Colorblind (Racial) Ideology**; entry for **Frames**).

Colorblind (Racial) Ideology This is a set of ideas that people use to rationalize or justify racial inequality in contemporary society without being labeled and stigmatized as racists (see entry for **Racial Ideology**; entry for **Ideology**; Bonilla-Silva 2014).

Commonsense Myths Ideas and narratives that do not represent social reality but become accepted as commonsense or taken for granted as true and therefore impact social reality.

Content Analysis A form of research that consists of looking for patterns and themes in the actual content of media communication.

Contested Social Issue A social issue that both receives a lot of public attention and is routinely the topic of debate or contestation. Which issues are contested and how they are contested is shaped by trends in society (see entry for **Debate Dynamics**; entry for **Social Problems**).

Controlling Images Representations of oppressed groups that associated them with features and characteristics that make their oppression seem like a natural outcome such as associating a group with aggression to explain higher rates of formal social control or laziness to explain higher rates of poverty or rationalize dismantling systems of economic support (see entry for **Culture of Poverty**; see entry for **Racialized Victim Blaming**; Collins 2008).

Counter-Framing Ways of understanding and explaining racially unequal outcomes or racialized events and issues that center the role of racial oppression and challenge the legitimacy of racial inequality (see entry for **Counterstories**; Feagin 2010).

Counterstories Narratives and forms of storytelling about events or issues that center the lived experiences, knowledge, and worldviews of oppressed groups such as people of color which contest the dominant stories told by groups with power (see entry for **Counter-Framing**; Deglado 2000; Solórzano and Yosso 2002).

Criminalization The process whereby groups, individuals, and activities become labeled as criminal. This is distinct from crime which is the violation of a law. For instance, not all instances of crime result in criminalization.

Critical Media Literacy The ability to engage with media content in ways that are mindful of the role of power and inequality in the content's messages, source of production, intended audience, and ways of framing social events and issues (see entry for **Framing**; entry for **Digital Racial Literacy**; Luke 1994).

Cultural Hegemony A means of maintaining authority via influence over mass culture in ways that help normalize and naturalize a group's dominant position and idealize their interests and perspective, which creates seemingly spontaneous and willing consent rather than coercion through force (see entry for **Hegemonic Whiteness**; Lears 1985).

Cultural Ideals Widely held ideas about the things that should happen or the things that people should be doing within a cultural context (see Weber 1958 [1904–1905]).

Cultural Racism A way of justifying racial inequality by claiming racially unequal outcomes are explained as solely the product of the cultural traits or cultural practices of people of color (see entry for **Culture of Poverty**; entry for **Colorblind (Racial) Ideology**; Bonilla-Silva 2014)

Culture of Poverty A theory that argues that many ethnic groups that remain in poverty do so because of the cultural traits and practices that they have adopted rather than social arrangements and opportunities, critically examined in the conclusion of Chapter 5 (see entry for **Cultural Racism**; entry for **Colorblind (Racial) Ideology**; entry for **Racialized Victim Blaming**; Steinberg 1989).

Debate Dynamics The trends, patterns, and overall social and cultural contours of a public debate about a contested social issue.

Decarceration A movement that seeks to reduce the jail and prison population, reduce the use of incarceration to address social problems, and reforming the legal system overall toward these ends as well as using policies that improve living standards and institutions of support to address the underlying causes of violence and victimization (see Drucker 2017).

Digital Divide Inequalities and disparities in who has access to digital technologies that are used for communication and accessing information (see Robinson et al. 2015).

Digital Media as a Social Practice As digital technology allows people to more actively participate in media, this perspective focuses on what people do with media in this context (see entry for **Media Studies**; Couldry 2012).

Digital Racial Literacy An understanding of both the specific cultural and technological context of online communication and digital media and its impacts on the dissemination of racial meanings and discourses (see entry for **Critical Media Literacy**; entry for **Racial Literacy**; Daniels 2008).

Discourse(s) Discourses are fairly coherent sets of categories, meanings, and stories that produce a picture of how the world works or a worldview. According to the philosopher Michel Foucault, discourses are used to produce knowledge and they are related to power dynamics in society.

Dominant Racial Meanings The ways that dominant groups understand and depict racial groups and categories in society (see entry for **Racial Meanings**).

Epistemology Epistemology is the ways in which people produce knowledge.

Epistemology of Ignorance Epistemology of ignorance describes the way whites or those of European background produce knowledge which ignores certain features of the social world, such as the causes and effects of racial oppression and the way that they differently impact people's lived experience and life chances (see entry for **Epistemology**; Mills 2007).

Folk Devil The characters or groups about which moral panics develop and become defined as threats to society (see entry for **Moral Panic**; Cohen 1972).

Four Aspects of Identity Major components of identity from a sociological perspective: (1) an ongoing process rather than something fixed, (2) constructed, in part, through language, (3) relates to choices people make and how society is organized, and (4) social categories and social groups.

Frames Ideas, concepts, and explanations that people employ in order to organize information and communicate. Frames, much like picture frames, allow people to emphasize or focus on particular perceived or real aspects of an issue to the exclusion of other aspects.

Framing Framing is the process of selecting and using frames to communicate in strategic ways about an event or issue (see entry for **Frames**).

The Fiscal Frame   A way of discussing a contested social issue that focuses on the financial implications such as how much something will cost, who will pay for it, or how much revenue could potentially be produced.

The Freedom and Equality Frame   A way of discussing contested social issues that focuses on abstract ideals about liberty and fairness in the discussion of contested social issues.

The Functionalist Frame   A way of discussing contested social issues that draws on the idea of functionalism or a focus on how well something achieves its stated purpose or goal. Within this frame, something may also be declared dysfunctional if it fails to achieve its assumed or stated purpose.

Group-Centered Frames   Ways of communicating that equate a policy or issue with a group of people so that people's responses about that issue or policy will be impacted by their feelings and perceptions about that group (see Hurwitz and Peffley 2005).

Hegemonic Whiteness   A racial identity rooted in the assumption that being white is a culturally normative state and a naturally dominant social position, including the way that people perform or act out the dominant image of what it means to be white which is idealized in contrast to negative conceptions about nonwhites or whites who fail to live up to this ideal (see entry for **Cultural Hegemony**; entry for **Identity Construction**; Lewis 2004; Hughey 2012).

Identity   Identity is one's sense of self or who they are (see entry for **Identity Construction**; entry for **Process of Identification**; entry for **Four Aspects of Identity**).

Identity Construction   The ongoing process of people negotiating, figuring out, and articulating their sense of self or who they are (see entry for **Process of Identification**).

Identity Opportunities   The set of subject-positions or possible perspectives that people have access to define themselves and others in ways that can be communicated and accepted by others (see entry for **Subject-Positions**; entry for **Identity**; entry for **Identity Construction**; entry for **Process of Identification**; entry for **Four Aspects of Identity**).

(Dominant) Ideology   The sets of ideas that are used to rationalize and justify the present state of society such as inequality or group domination (see entry for **Racial Ideology**;

entry for **Colorblind (Racial) Ideology**; Marx and Engels 1978[1845–1846]).

Inequality An unequal distribution of symbolic and material rewards among groups in a society (see entry for **Racial Inequality**).

Institutional Racism How the rules and routine practices of social institutions produce racial disparities in resources and opportunities such as laws targeting certain groups or disproportionately impacting them (see entry for **Racial Oppression**; entry for **Racial Inequality**; Ture and Hamilton 1967[1992]).

Just-World Fallacy A fallacious worldview or way of thinking which assumes that all outcomes are inherently just, and therefore negative and positive situations that people find themselves in are entirely deserved or brought about by their own actions or characteristics (see entry for **Ideology**; entry for **Racial Ideology**; entry for **Colorblind (Racial) Ideology**; entry for **Racialized Victim Blaming**; Lerner 1980).

Ku Klux Klan A white supremacist terrorist organization originally founded by Confederate soldiers after the Civil War in 1866 in Tennessee that experienced a resurgence in the 1920s and remains one of the most influential hate groups in the United States.

Mass Media The means through which information, images, and other messages are distributed by corporate entities, including the Internet, television, film, radio, and print media such as magazines, literature, and newspapers.

Media Consolidation A process whereby fewer and fewer companies produce media content as media corporations merge and smaller producers are bought up.

Media Studies A broad field that analyzes many aspects of media, including its social context and connection to social processes and the production, distribution, consumption, and effects of media and media practices (see Hall 1980).

Moral Categories The concepts that people use to position things as either moral or immoral (or good or evil, sacred or profane, innocent or guilty, etc.) in everyday life that often rely on binaries which influence our ability to empathize and the scope and trends of debates over contested social issues.

Moral Entrepreneurs Individuals in society who engage in sustained efforts to construct rules and have them enforced (see Becker 1964).

Moral Panic A type of social process whereby a group, activity, or event is defined as a threat and therefore regarded with fear and anxiety by sectors of the population (see entry for **Folk Devil**; Cohen 1972).

Myths The ways that language and power can present certain narratives and ideas as natural or a given, often operating through appeals to commonsense and remove all historical context to make things that take place through a social process appear as though they are simply natural (see entry for **Commonsense Myths**; Barthes 1957)

Ontological Investment A sense of allegiance or sunken cost when people define themselves and the world around them using a particular set of categories or ideas (see entry for **Ontology**; entry for **Ontological Security**).

Ontological Security A sense of continuity and comfort in the flow and meaning of events in one's life, expectations about routines, habits, and patterns in terms of what exists or can be expected (see entry for **Ontology**; entry for **Ontological Investment**; Giddens 1991).

Ontology Inquiry which is concerned with issues of being or what exists (see entry for **Ontological Investment**; entry for **Ontological Security**).

Oppression A set of social relations between groups where one group holds more power and influence and uses that power and influence to disadvantage other groups (see entry for **Racial Oppression**).

Othering The creation of a distinction between self and "other," often with an emphasis on privileging the self over the other and making claims about the "other" as homogenous, not normal or ideal (see entry for **Symbolic Boundaries**).

Primary Institutional Definers People who represent major institutions or organizations or who are considered experts that have disproportionate influence over how issues and events are defined in the media. Media organizations often rely on these definers so as to appear neutral or objective (see Hall et al. 1978).

Process of Identification A process whereby people identify with certain discourses and subject-positions made available to them by institutions and other people (see entry for **Discourse**; entry for **Identity Construction**; entry for **Subject-Position**)

Public Sphere A space in which people can discuss and deliberate in order to form a consensus about the best appropriate social or political action on a particular issue (see Habermas (1964[1974]).

Racial Apathy Attitudes that whites hold about racial inequality where they may recognize but do not care about this issue or see it as a priority (see Forman 2006).

Racial Empathy Gap Whites' common inability to empathize with, or imagine and feel for the lived experiences of, people of color (see Silverstein 2013; Bonilla-Silva 2016).

Racial Ideology Sets of ideas that people use to interpret racially unequal outcomes in society as just, natural, or reasonable (see entry for **Ideology**; entry for **Colorblind (Racial) Ideology**; Bonilla-Silva 2014).

Racial Inequality The unequal distribution of material (i.e., money and property) and symbolic (i.e., respect and prestige) resources in society along the lines of groups formed around racial categories. Racial inequality is thus an artifact of how society is arranged and what people do in everyday life rather than essential group traits or natural causes. Racial inequality benefits those socially ascribed to the racial category of "white."

Racial Justice The ideal that people should receive fair treatment regardless of their racial categorization. As many activists and scholars have noted, substantial progress and social change, including the redistribution of resources and opportunities and transformations, in dominant cultural meanings are required for the achievement of racial justice.

Racial Literacy An aptitude to use a vocabulary, set of concepts, and historical knowledge to perceive, understand, and discuss the role of racial categories and racial oppression in society and people's everyday lives (see entry for **Racial Silence**; entry for **Digital Racial Literacy**; Twine 2010).

Racial Meanings How racial categories, structures, and groups are culturally represented in a society (see entry for **Dominant Racial Meanings**).

Racial Oppression A social relationship between racial groups in which one group ("whites") maintain advantage and power over other groups (see entry for **Oppression**; entry for **Racial Inequality**).

Racial Silence An overall trend within a particular social setting or conversation of a racialized topic where its racial implications are omitted, results from norms that racial issues and language are unacceptable, the assumption that discussion of racial issues or language are not advantageous, or the mass avoidance of knowledge about the racial implications of an issue or event (see entry for **Racial Literacy**; Rosino and Hughey 2017)

The Racial Unfairness Frame Similar to the ideal of racial justice, a way of discussing a contested social issue which focuses on ways that people receive unjust treatment or encounter bias on the basis of their placement in a racial category.

Racialization The process whereby things and people (i.e., institutions, places, social practices, public policies, ideas, physical traits, etc.) are given racial meanings or associated with racial categories (see Omi and Winant 1994).

Racialized Criminal Threat Claims about crime that specifically tie the threat of crime or other forms of social threat to specific racial groups (see entry for **Moral Panic**; entry for **Racialization**; entry for **Code Words**).

Racialized System of Social Control Organized and formal means of controlling or regulating populations such as incarceration or policing which employ dominant racial meanings and help maintain racial structures of inequality and oppression (see entry for **The War on Drugs**; entry for **Racialization**; entry for **(Dominant) Racial Meanings**; Provine 2007; Alexander 2012).

Racialized Victim-Blaming The act of finding ways in which the victims of inequality are to blame or have faults through the use of racial meanings or the way that racial categories are defined (see entry for **Racialization**; entry for **(Dominant) Racial Meanings**; Ryan 1976).

Racism A system (including beliefs, practices, and policies) that combines to reproduce and maintain racial oppression and inequality (see entry for **Racial Oppression**; entry for **Racial Inequality**).

Resonance The extent to which a cultural object aligns with the interests and worldview of its audience or consumer, it is said to 'resonate' (see Schudson 1989; Rosino and Hughey 2017).

Scientific Racism The use of scientific concepts such as evolution, genetics, or biology to attempt to rationalize or naturalize racial inequality or justify systems of racial oppression. This mode of interpreting racial categories and racially unequal outcomes has been debunked but remains disturbingly prevalent in many contexts (see entry for **Ideology**; entry for **Racial Ideology**; entry for **(Dominant) Racial Meanings**).

Social Disorganization A theory advanced by social scientists in the 20th century that argues urban spaces do not have a clear sense of shared norms and social bonds and therefore is more conducive to crime and the use of formal means of crime control such as prisons. This view has severe limitations, as discussed in Chapter 2.

Social Information Information (such as that provided by mass media) that people use to form opinions on social issues, a sense of who they are, and ideas about their place in the social world (see Entman and Rojecki 2001).

Social Position The relative placement of a social group within societies characterized by inequality or a hierarchy, which reflects things such as relationships to other groups (including conflict, solidarity, and oppression) and relationships to institutions that provide access to resources (see Collins 1993).

Social Problem An issue or event that has been collectively identified as a problem for society or specific social groups (see entry for **Contested Social Issue**; Blumer 1971).

Sociological Imagination The ability to draw connections between individual people's troubles and large-scale public issues such as historical or social trends or major events within the context of society (see entry for **Sociological Knowledge**; entry for **Sociological Mindfulness**; Mills 1959).

Sociological Knowledge Knowledge about one's self, others, and society that are gained through empirical analysis of social events and issues and the cultivation of sociological imagination to interpret trends and dynamics (see entry for **Sociological Imagination**; entry for **Sociological Mindfulness**).

Sociological Mindfulness The ability to take insights drawn from making sociological connections and taking them into consideration in everyday life (see entry for **Sociological Imagination**; entry for **Sociological Mindfulness**; Schwalbe 2005).

Sociological Research The systematic collection and analysis of data to answer questions about society and social life. It can focus on small- or large-scale examinations of events, issues, and processes.

Structural Change A constant ongoing transformation in how rewards, punishments, and opportunities are distributed among groups and individuals and alteration to the informal and formal rules that govern how we interact with each other; this process is impacted by collective action (see entry for **Structure**; entry for **Collective Action**; entry for **Structural Racism**).

Structural Racism The ways that racial inequality is produced by social structure or the way that society is organized in terms of rules and the distribution of resources and penalties. This can include multiple institutions or aspects of society (see entry for **Institutional Racism; Racial Oppression**; entry for **Racial Inequality**; Bonilla-Silva 1997).

(Social) Structure The way that society is organized, including how resources are distributed, what form and informal rules exist and how they are enforced, and the routines and patterns of people's everyday lives and interactions (see entry for **Agency**; entry for **Structural Change**; Giddens 1993).

Subject-Position(s) Sets of categories, stories, and ideas (discourses) that people can identify with to form an identity. In a way, people are subjects or actors and observers of the world around them (in the sense of subjectivity), and these describe the positions from which people are able to do so based on the discourses they encounter and use (see entry for **Process of Identification**; entry for **Identity**; entry for **Discourse**; entry for **Identity Construction**; Hall 1996, 1997).

Symbolic Boundaries The distinctions that people make to categorize people, places, things, and other aspects of the world around them as different. Symbolic boundaries also describe the logic or meanings that people apply in making such distinctions (see entry for **Moral Categories**; Lamont 2000).

Symbolic Power The power that dominant groups hold to impose their language, worldview, ideas, and expectations onto the social world with legitimacy and authority, the ability to define reality and impose that definition on others, often backed by powerful institutions such as the government (see Bourdieu 1991).

The Vietnam War An international conflict from 1555–1975 between the US military and resistance forces loosely allied with Communist China in Vietnam, Cambodia, and Laos. Despite being lumped in with the larger geopolitical conflict known as the "Cold War," which implies a lack of bloodshed, estimates vary that the Vietnam War resulted in the deaths of between 700,000 and 4,000,000 civilians and military members. It also included numerous war crimes and the use of chemical weapons by the US military. The war also sparked waves of mass resistance in the United States among drafted service members and civilians (see MacLear 1982; Cortright 2005; McCormick 2000).

The War on Drugs A phrase describing the strategy of enforcing drug laws which relies on imprisonment, aggressive policing and the use of force, and routine targeting of marginalized groups such as impoverished black and Latino men. It originally derives from the use of war-like language in Richard Nixon's 1971 speech on drug addiction.

The War on Drugs Debate Public discussion, argument, and contestation over the purpose and consequences of the War on Drugs and drug policies enforcement more broadly. This debate primarily takes place through mass media (see Rosino and Hughey 2016).

Whiteness Studies An interdisciplinary field that seeks to make the relationships between racial oppression and whiteness (or the category of white) visible and analyze how meanings of race and practices of racism impact those who are racialized as white in their everyday lives (see Twine and Gallagher 2008).

Index

Note: **Bold** page numbers refer to tables and page numbers followed by "n" denote endnotes.

"abstract liberalism" 112
affect 36
agency 13
agenda-setting 17, 43–45
alcohol 6, 57, 98; prohibition 6, 7, 9, 99
Alexander, Michelle 33; *The New Jim Crow* 11, 49
Althusser, Louis 120n3
American Journal of Bioethics 11
American Protestants 6
Anslinger, Harry J. 7, 39
anti-black police violence 1
anti-black violence 96
Anti-Catholic sentiment 6
anti-drug rhetoric 33
AT&T 24
audiences 85–88

Bagdikian, Ben H.: *The Media Monopoly* 23
Balko, Radley 33; *The Rise of the Warrior Cop* 49
Barthes, Roland 128
Baum, Dan 2
Becker, Howard S. 4, 31
Beckett, Katherine 10, 43, 44
Berensen, Alex: *Tell Your Children: The Truth About Marijuana, Mental Illness, and Violence* 11–12

Berger, Peter L. 91, 134
Biden, Joseph 35
black families 105, 106, 115
black people 24, 31, 34, 104, 117, 128, 130, 132–133, 141
black press 86, 87
Blumer, Herbert 12, 29
Bonilla-Silva, Eduardo 72, 78, 87, 108, 112
Bourdieu, Pierre 137
Buckley, Charles F. 42
Buckley, William F. 42, 43
Bush, George W. 34, 49

cannabis/marijuana 1, 2, 7–9, 12, 35, 39, 48, 96, 98, 99, 139–141
Catholicism 6
The Chicago Tribune 3
Chinese Exclusion Act (1882) 5
Chinese immigrants 4, 5, 8
Civil Rights era 76, 79
Civil Rights Movement 9, 23, 42, 131
Civil War 6
Clinton, Bill 80
Clinton, William J. 35
cocaine 7–9, 34, 41n3, 111
code words 18, 77–81, 116–117, 129
collective action 20
collective definitions 12, 29, 32, 91

Collins, Patricia Hill 28, 136
colorblind defense frame 110–113
colorblind ideology 72, 124
colorblind victim blaming **94**
color evasiveness 67
Comcast 24
comments: online 93–94, **94**, 117, 121n11, 128; on War on Drugs articles 102
comment writers 98
commonsense myths 19, 127–131
communication 86, 88, 90
conflict, racial and political 38
contemporary mass media 30
content analysis 16
contested social issue 17, 19, 21, 29, 32, 44, 64, 67, 73, 75, 81, 85, 88, 90, 92, 93, 102, 114, 116, 122, 127, 131, 134, 135, 137–139, 141
controlling images 28, 133, 141
Cooley, Charles Horton 120n8
Couldry, Nick 30
counter-framing 118
counterstories 118
crack 54
"crack baby" myth 111, 116
"crack epidemic" 9, 46, 111
crime: debates over 25–29; in mass media 74; moral panics about 33
crime rates 26, 27, 29, 44
criminality 105
criminalization 25, 36, 115, 118, 126, 138
criminal justice reforms 38
criminal justice system 124; intrinsic fairness of 109–110; racial inequality in 107; racially disproportionate outcomes in 108
criminal stereotypes 82n1
criminology 10
The Crisis 86
critical media literacy 138
cultural hegemony 124–125
cultural ideals 76–77
cultural racism 105, 124
culture of poverty 115

Daniels, Jessie 138
Dark Alliance (Webb) 11
debate: about racism 92; mass media coverage of 114; in media 66; in newspapers 128; over contested social issues 93; over race, crime, and immigration 25–29; *see also* War on Drugs debate
debate dynamics 81
decarceration 141
Delgado, Richard 118
Desrochers, Robert E., Jr. 86
The Detroit News 87
Detroit race riot (1943) 87
deviance 107
digital divide 84
digital inequalities 84
digital media 2, 23, 84, 86, 89, 117, 123, 137, 138; platforms 85; as social practice 85
digital racial literacy 138
digital spaces 84, 88
digital technologies 84
discourse 91, 95, 102, 118; media 114; political 80; public 80, 130; racial 91, 92, 114–115, 117
discrimination 2, 28, 71, 101, 109; racial 10–11, 141
divergent identities 134
Doane, Ashley W. 70, 82n2
dominant group 31, 73, 77, 106, 115, 135, 137, 138
dominant ideology 123–124
dominant racial ideology 124, 126
dominant racial meanings 19, 123, 126
drug addiction, as public health problem 56
drug crimes: nature and qualities of 61; in United States 59
drug epidemics 37
drug law enforcement 2, 10, 31, 32, 34, 97, 128
drug laws: development and implementation of 6; militarization and racialization of 32–40

drug policy 31, 38; controversial social issues of 16; failure of 53; harshness of 2; racial silence in 73; televised debates over 42; in United States 8
Drug Policy Alliance 83n4
drug policy enforcement, militarization of 37
"drug problem" 34, 35
drug prohibition: history of 4; policies and practices 9–10
drug prohibition policies, motivations and consequences of 2
drug use 37, 44
Du Bois, William Edward Burghardt 26–28, 86, 106, 121n9, 131
Durkheim, Emile 53, 120n6

Eastland, James O. 9
economic class inequality 97
Ellwood, Charles A. 27, 28
Elwood, William N. 41n3
Embrick, David G. 87
empathy 131–133
The Emperor Wears No Clothes (Herer) 11
Engels, Freidrich 72
Entman, Robert M. 88, 117
epistemology 68
epistemology of ignorance 68
Essed, Philomena 36
ethnic groups 115
European liberal democratic societies 22
European migrants 6, 28

families: male-headed 116; relationship between societies and 106
Feagin, Joe R. 118
Federal Bureau of Narcotics 7
Fellner, Jamie 32
Firing Line (1991) 42
fiscal frame 18, 47–48, 95; freedom and justice frame 48–52; functionalist frame 52–59

folk devil 8
four aspects of identity 88
The Fox Corporation 24
frames/framing 17; in comments 94, **94**; debate 43–45; fiscal frame 95; freedom and justice frame 95–98; functionalist frame 98–100; racial unfairness frame 100–102
Frank, Deborah A. 111
Frankenberg, Ruth 67, 69
Fraser, Nancy 22
freedom and justice frame 18, 48–52, 95–98
Freedom's Journal 86
Free Speech and Headlight 86
functionalist frame 18, 52–59, 64, 98–100

Gaytán, Marie Sarita 6
Giddens, Anthony 126
Glickman, Lawrence B. 75
Goffman, Erving 44, 90, 120n1
Graham, Ramarley 1–3
Gramsci, Antonio 124–125
group-centered frames 45
The Guardian 3

Habermas, Jurgen 21–22
Hall, Stuart 7, 31, 75, 92, 134
Hart, Carl: *High Price* 11
hegemonic whiteness 125
Herer, Jack: *The Emperor Wears No Clothes* 11
heroin 3, 4, 9, 37
High Price (Hart) 11
Hoffman, Frederick L. 27
Hoffman, Seymour 37
Holder, Eric 38, 51
Hourwich, Isaac A. 27
human rights 49, 142–143
Hurwitz, Jon 45, 64, 80

identity 13, 14, 134–137; commenting and constructing 88–93; four aspects of 88; racial meanings from ideology to 123–127

identity construction 90, 91, 117, 119; morality in 134
identity opportunities 136
identity politics 135
ideology 72, 90; to identity, racial meanings from 123–127
immigration 17, 21, 29, 58, 80; debates over 25–29
imprisonment 25
in-depth ethnographic research 25
inequalities 82n1
inequality 10, 12–13, 15, 26, 115, 136; economic class 97
institutional racism 70–71, 114
internal colonialism 71
Internet comments 16, 19, 84, 94, 95, 129
interpretations, of crime statistics 28
investment, of public funds 3

Jackson, Andrew 79
Jacobs, Ronald N. 31
Jefferson, Thomas 79
Jewish immigrants 6
Jim Crow 28, 140
journalism 30
Judaism 6
just-world fallacy 126

Kadowaki, Joy 39
Kapernick, Colin 44
Kelly, Brian C. 39
King, Martin Luther 131
Ku Klux Klan 6, 79, 87

Latinx people 24, 31, 34, 104, 117, 128, 130, 132–133, 141
law enforcement 38; racialization and militarization of 34
legalization 98, 140
LePage, Paul 129, 130
Lewis, Amanda E. 68, 82n2, 124
Lewis, Oscar 115
Lexis Nexus 46
Lopez, German 12
Lopez, Ian Haney 78
Los Angeles Herald 4–5
Luckmann, Thomas 91, 134

Malcolm X 133
male-headed families 116
Marihuana Tax Act 7
Marx, Karl 72
mass communication 24
mass imprisonment 48
mass incarceration/overcrowding 2, 21, 29, 31, 36, 50, 51, 77, 97
mass media 23, 29, 75, 127, 135, 137, 138; crime in 74; forms of 42; racialized group representation in 36; War on Drugs critics in 54
mass media coverage 114
mass media messages 70
Mead, George Herbert 120n4
Meads, Mallory 33
media 85–88; agenda setting and debate framing 43–45; debates in 66; implications for 137–143; role of 29
media-audience relationships 86, 88
media audiences 85, 88
media communication 16
media consolidation 23–25
media content 87, 117; about War on Drugs 90; robust relationship between 88
media debates, elites and power dynamics of 30–32
media literacy 143
The Media Monopoly (Bagdikian) 23
media organizations 74
media ownership 24
media technologies 84
Mendelberg, Tali 80
methamphetamine 39
Mexican immigrants 6–8
Miami Herald 78
Michigan Chronicle 87
militarization 3, 17, 49, 141; of drug laws 32–40
Military Cooperation with Law Enforcement Act (1981) 33
military equipment 96
Mill, John Stuart 51
Mitchum, Kennedy 92

moral categories 133
moral enterprise 4, 6
moral enterprise/entrepreneur 4, 6–9, 11, 39
morality 133–134
moral panic 7–9, 11, 12, 18, 33, 66, 73, 100, 127, 128
Moynihan, Daniel Patrick 115, 116
Mueller, Jennifer C. 68
Muhammad, Khalil Gibran 28
Murphey, Lynne 103
Muslim extremists 100
myths 127–131

narcotic substances 7
National Commission on Marihuana and Drug Abuse 8
National Review 42
"Negro cocaine fiend" 8
"the Negro question" 27
The New Jim Crow (Alexander) 11, 49
news media 31
newspaper articles, frames and themes in **46**, 46–47
newspapers 74–76, 86–87, 95, 117, 129; debate in 128; on War on Drugs 103
New York Academy of Medicine 8
New York Police Department (NYPD) 1
The New York Times 12, 30, 89
Nietzsche, Friedrich Wilhelm 133
Nightingale, Virginia 85
Nixon, Richard 3, 8, 42, 62, 79, 98
Northpointe 38
Nyrop, Kris 10

Obama, Barack 39, 80, 88
The Onion 53
online comments 117, 121n11, 128; analyzing of 93–94, **94**
online news 88
online spaces 84–85, 138
ontological investment 126
ontological security 126
ontology 126
opioids 37
opium 4–5

opium bans 5
oppression 15
othering 103

Pager, Devah 11
Parsons, Talcott 53
Partnership for a Drug-Free America 34
Pascale, Celine-Marie 131
Peffley, Mark 45, 80
personal responsibility 112, 113, 116, 124
Pfingst, Lori 10
pleas, for sanity 99
police brutality 44, 96
policing 2, 10, 11, 25–27, 30, 33, 34, 36–38, 40, 47, 60, 71, 111, 141; implications for 137–143
political conflict 38
political identity 134, 142
post-Civil Rights era 35
poverty 26, 39, 80, 106, 116
power dynamics, of media debates 30–32
power evasiveness 67
primary institutional definers 31
print media 86, 123; War on Drugs in 45–47, **46**
prison overcrowding 50
prison system 9
process of identification 90
Protestant ethic 77
public debates 2, 3, 12, 15, 24, 27, 29–31, 43, 59, 80, 81, 84, 85, 93, 127, 129, 130, 132, 139, 142
public funds 3
public health-oriented approaches 56
public opinion research 29
public resources 2
public sphere 17, 21–23, 29
punitive drug policies 66

race/racism 4, 6, 15, 36, 46, 59, 70, 72, 76, 78, 85, 91–92, 101, 108; cultural 105; debate about 92; debates over 25–29; in form of negative attitudes 109; institutional 70–71, 114; moral

panics about 33; multiracial heritage for 67; scientific 26; in United States 67
racial apathy 114
racial bias 70; in War on Drugs 71–72
racial categories 92–93
racial code words 80
racial conflict 38
racial discourse 91, 92, 114–115, 117
racial discrimination 10–11, 141
racial disparity 70, 103, 107, 123, 124, 128; causes and consequences of 129; in criminalization 36; in War on Drugs 106
racial dynamics 108
racial empathy gap 132, 133
racial equality 42, 118
racial group 45, 66, 77, 93, 103
racial identity 12–13, 21, 123, 125, 129, 134
racial identity construction 19
racial ideology 72, 123, 124
racial inequality 2, 15, 16, 21, 36, 46, 71, 73, 77, 88, 93, 97, 101, 114, 116, 123, 129; in criminal justice system 107; in United States 67
racial injustice 44, 116, 119, 124
racialization 66; of drug laws 32–40
racialized criminal threat 99
racialized group, representation in mass media 36
racialized mass incarceration 39, 112, 140, 141
racialized moral panic 8, 62, 127, 134
racialized social system 15, 59, 68, 71
racialized subject-positions 126, 134, 135
racialized system of social control 39
racialized threat 100
racialized victim-blaming frame 103–110

racial justice 48, 67, 73–74, 81, 114, 117, 118, 128, 133, 143
racial literacy 67, 131
racial meaning, from ideology to identity 123–127
racial oppression 15, 26, 29, 31, 36, 39, 68, 85–88, 102, 121n9, 141; hegemonic whiteness 125; in United States 128
racial silence 18, 67–74, 81, 117, 118; in drug policy 73; emergence of 72; in War on Drugs debate 69
racial stereotypes 34, 88
racial terrorism 86, 87
racial trait 27, 29, 58, 104
racial unfairness 71, 101, 112
racial unfairness frame 18, 59–63, 100–102
racist tropes 107
Rangel, Charles 42, 43
Reagan (President) 33, 34
"Reefer Madness" campaign 12
resonance 18, 74–77, 81, 119
The Rise of the Warrior Cop (Balko) 49
risk assessment tools 38
Robins, Kevin 24
Robinson, Matthew 138
Rojecki, Andrew 88, 117
Room for Debate 30
Rosen, Jay 85
Ross, Karen 85
Ryan, William J. 103

Sánchez-Jankowski, Martín 25
San Francisco's Cannabis Equity Program 141
Schudson, Michael 76
Scientific American 115
scientific racism 26, 121n9
'serious violent felony' 35–36
Sessions III, Jefferson Beauregard 38
sexism 85
Sinclair Broadcast Group 24
small-scale media 24
social control 25, 26, 59, 95, 110, 128, 142; racialized system of 39
Social Darwinism 105

Index   187

social disorganization 25, 26
social forces 85
social groups 13, 29
social identity 142
social inequality 26, 135, 140
social information 87
social institutions 127, 135
social media 23
social movements 142
social outcomes 93
social position 68
social problem 12, 29, 38, 43, 63, 118, 122, 133–135, 137, 139–142
social structure 14, 91, 101, 123, 134
sociological imagination 13, 14, 112
sociological knowledge 136
sociological mindfulness 14
sociological research 16
spaces: digital 84, 88; online 84–85, 138
Steinberg, Stephen 68, 105, 116
stereotypes 82n1, 107, 117
structural change 140–142; implications for 137–143
structural racism 101
structure 13
subject-position 19, 82, 92, 95, 104, 126
subordinate racial groups 135
Sullum, Jacob 51
SWAT raids 33
symbolic boundaries 19, 92, 95
symbolic power 137
Syrian refugees 83n6
systemic racism 15, 59, 71, 104, 118
Szoldra, Paul 37

tax revenue 2, 48
television programming 42
Tell Your Children: The Truth About Marijuana, Mental Illness, and Violence (Berensen) 11–12
temperance movements 6
theme: in comments 94, 94; focused on prison-industrial complex 97

"Three Strikes Law" 35
Townsley, Eleanor 31
Trump, Donald J. 38, 39, 78, 83n6, 88
Twine, France Winddance 67
two tiered approach to drug policy 35

unique frames 102–103; colorblind defense frame 110–113; racialized victim-blaming frame 103–110
United Kingdom 4
United States 1; birth of mass media in 86; black families in 115; crime rates in 29; digital divide 84; drug crimes in 10, 59; drug policy in 8; drug prohibition history in 4; drug prohibition laws in 141; prison system in 9; public sphere in 22; racial oppression in 128; racism and racial inequality in 67
urbanization 25, 26
urban policing 26

ViacomCBS 24
victim-blaming 103, 112, 115, 124, 130
Vietnam War 3
violent crime 38, 45
Violent Crime Control and Law Enforcement Act (1994) 35–36
Vuolo, Mike 39

Wacquant, Loïc 39
Wallace, George 79
The Walt Disney Company 24
"war against heroin addiction" 3
War on Drugs 2–11, 29; claims of 47, 96; criticism of 99; critics in mass media 54; debate about 66; drug prohibition policies and practices 9–10; "drugs" and "medicine" 4; dysfunction of 59, 64, 99; economic implications of 48; economic opportunity costs of 95; human costs of 52; manifest functions 53;

media content about 90; moral entrepreneurs 6–7; moral panic 7–8; newspapers on 103; opium bans 5; opposition to 42; police brutality against residents 34; policies and practices of 51, 57, 100, 107, 112, 127; in print media 45–47, **46**; public debates about 43; public funds 3; racial bias in 71–72; racial discrimination 10–11; racial disparity in 106; racial implications of 61–62; racialized mass incarceration 39; racially unequal outcomes of 112; unfairness of 101

War on Drugs debate 2, 11–15; racial silence in 69; understanding through research and data 16–17; *see also* debate

warrior-style policing 141

Washington, George 79

Webb, Gary: *Dark Alliance* 11

Weber, Max 77

Weekly Butte Record 5

Wells, Ida B. 86

Wellstone, Paul 23–24

whiteness 125, 126; innocence of 104; invention of 69; studies 69

white supremacy 77

Wilson, Thomas Woodrow 79

"yellow menace" 100

Zerubavel, Eviatar 73